Elliptic Partial Differential Equations with Python

Jamie Flux

https://www.linkedin.com/company/golden-dawn-engineering/

Contents

1 Laplace's Equation   **10**
    Historical Background and Importance 10
    Mathematical Formulation 10
    Fundamental Solutions 11
    Boundary Value Problems 11
    Applications in Electrostatics 11
    Python Code Snippet 12
    Multiple Choice Questions 14

2 Poisson's Equation   **17**
    Derivation and Physical Context 17
    Green's Functions 18
    Numerical Methods: Finite Difference 18
    Boundary Conditions 18
    Applications in Heat Transfer 18
    Python Code Snippet 19
    Multiple Choice Questions 21

3 The Helmholtz Equation   **24**
    Mathematical Setup and Physical Significance 24
    Separation of Variables 25
    Methods of Fundamental Solutions 25
    Eigenvalue Problems 25
    Applications in Acoustics 26
    Python Code Snippet 26
    Multiple Choice Questions 28

4 The Biharmonic Equation   **31**
    Governing Equations in 2D and 3D 31
    Boundary Conditions and Their Types 32

    Green's Function Approach 32
    Numerical Implementation 33
    Applications in Elasticity Theory 33
    1   Example Application: Bending of a Thin Beam 34
    Conclusion 34
    Python Code Snippet 34
    Multiple Choice Questions 36

5 The Diffusion Equation   39
    Mathematical Derivation 39
    Analytical Solutions 40
    Numerical Methods: Finite Element 41
    Stability and Convergence 41
    Applications in Biological Processes 42
    Summary 42
    Python Code Snippet 42
    Multiple Choice Questions 44

6 Navier-Stokes Equations for Incompressible Flow   47
    Fundamental Principles 47
    Derivation and Physical Context 48
    Weak Formulation 48
    Vorticity-Stream Function Approach 48
    Computational Fluid Dynamics Methods 49
    Real-World Applications 49
    Python Code Snippet 49
    Multiple Choice Questions 52

7 Maxwell's Equations in Electromagnetics   55
    System of Equations and Basic Concepts 55
    Time-Harmonic Case 56
    Boundary-Integral Methods 57
    Finite Element Methods 57
    Applications in Telecommunications 57
    Python Code Snippet 58
    Multiple Choice Questions 60

8 The Cauchy-Riemann Equations   63
    Introduction to Complex Analysis 63
    1   Representation in the Complex Plane 64
    2   Analytic Functions 64
    Mathematical Formulation 64

    Polar Coordinates Solutions 64
    Numerical Implementations 65
    Applications in Fluid Dynamics 65
    Python Code Snippet 66
    Multiple Choice Questions 68

9 Schrödinger Equation   71
    Quantum Mechanics Background 71
    Mathematical Formulation 71
    Analytical Solutions for Simple Potentials 72
    Spectral Methods 72
    Solving Time-Dependent Schrödinger Equation . . . 73
    Quantum Computing Applications 73
    Python Code Snippet 74
    Multiple Choice Questions 76

10 The Monge-Ampère Equation   79
    Geometric Background 79
    Weak Solutions 79
    Numerical Solutions via Finite Differences 80
    Convex Optimization Algorithms 80
    Applications in Geometric Optics 80
    Python Code Snippet 81
    Multiple Choice Questions 83

11 The Black-Scholes Equation   86
    Introduction 86
    Financial Derivatives and Option Pricing 86
    Derivation and Assumptions 87
    Analytical and Semi-Analytical Solutions 87
    1   Closed-Form Solutions 87
    2   Numerical Solutions 88
    Practical Applications in Finance 88
    Python Code Snippet 89
    Multiple Choice Questions 92

12 Elliptic Regularity Theory   95
    Sobolev Spaces 95
    1   Definition 95
    2   Norm and Completeness 96
    3   Sobolev Embedding Theorems 96
    Weak Solutions & Lax-Milgram Theorem 97

	1	Definition of Weak Solutions 97
	2	Lax-Milgram Theorem 97
	3	Applications 98

Schauder Estimates 98
    1   Statement of Schauder Estimates 99
    2   Practical Significance 99
    3   Extensions and Refinements 100
Python Code Snippet 100
Multiple Choice Questions 103

13 The KdV Equation   106

Introduction to Solitons 106
    1   Types of Solitons 106
    2   The KdV Equation in Physical Context . . . 107
Method of Inverse Scattering Transform 107
    1   Scattering Problem 107
    2   Construction of Solitons 107
Numerical Methods 108
    1   Finite Difference Methods 108
    2   Spectral Methods 108
    3   Pseudospectral Methods 108
Analytical Solutions 109
    1   Solitary Wave Solutions 109
    2   Periodic Wave Solutions 109
    3   Rational Solutions 109
Applications 109
Conclusion 110
Python Code Snippet 110
Multiple Choice Questions 113

14 The Ricci Flow Equation   115

Geometric Analysis Background 115
Hamilton's Formulation 115
Long-Time Behavior 116
Numerical Simulations 116
Applications in Geometry and Topology 116
Python Code Snippet 116
Multiple Choice Questions 118

15 Variational Methods in Elliptic PDEs — 121
- Calculus of Variations 121
- Euler-Lagrange Equations 121
- Functional Spaces 121
- Minimization Techniques 122
- Applications in Physics and Engineering 122
- Python Code Snippet 122
- Multiple Choice Questions 125

16 The Heat Equation on Manifolds — 128
- Differential Geometry Basics 128
- Formulation on Riemannian Manifolds 128
- Spectral Properties 129
- Numerical Approaches 129
- Applications in Geometry 129
- Python Code Snippet 130
- Multiple Choice Questions 131

17 The Allen-Cahn Equation — 134
- Phase Field Models 134
- Analytical Solutions 134
- Numerical Simulations 135
- Application to Material Science 135
- Pattern Formation and Dynamics 135
- Python Code Snippet 136
- Multiple Choice Questions 137

18 The Ginzburg-Landau Equations — 140
- Superconductivity Theories 140
- Mathematical Formulation 140
- Vortex Solutions 141
- Numerical Techniques 141
- Applications in Condensed Matter Physics 141
- Python Code Snippet 142
- Multiple Choice Questions 144

19 Elliptic Integro-Differential Equations — 147
- Introduction and Basic Examples 147
- Analytical Techniques 148
- Probabilistic Interpretation 148
- Numerical Methods 148
- Applications in Financial Mathematics 148

Python Code Snippet 149
　　　Multiple Choice Questions 151

20 The Fokker-Planck Equation　　　　　　　　　154
　　　Stochastic Processes Background 154
　　　Derivation and Analytical Solutions 155
　　　Numerical Methods 155
　　　Long-Time Behavior and Stationarity 155
　　　Applications in Statistical Mechanics 155
　　　Python Code Snippet 156
　　　Multiple Choice Questions 158

21 The Fisher-KPP Equation　　　　　　　　　　　161
　　　Population Genetics Context 161
　　　Traveling Wave Solutions 161
　　　Analytical Techniques 162
　　　Numerical Simulations 162
　　　Applications in Epidemiology 162
　　　Python Code Snippet 163
　　　Multiple Choice Questions 165

22 The Fractional Laplacian　　　　　　　　　　　168
　　　Definition and Properties 168
　　　Spectral Representation 169
　　　Numerical Approximations 169
　　　Boundary Value Problems 169
　　　Applications in Anomalous Diffusion 170
　　　Python Code Snippet 170
　　　Multiple Choice Questions 172

23 Elliptic Systems of Equations　　　　　　　　　175
　　　Coupled PDE Systems 175
　　　Operator Theory 176
　　　Sobolev Spaces for Systems 176
　　　Numerical Methods 176
　　　Applications in Multiphysics Problems 176
　　　Python Code Snippet 177
　　　Multiple Choice Questions 179

24 The Lichnerowicz Equation — 182
- Relativity Theory Context 182
- Conformal Metrics 182
- Analytical and Numerical Solutions 183
- Initial Data in Relativity 183
- Applications in Cosmology 183
- Python Code Snippet 184
- Multiple Choice Questions 185

25 The Obstacle Problem — 188
- Variational Inequality Formulation 188
- Free Boundary Problems 188
- Regularity Theory 189
- Numerical Solutions 189
- Applications in Finance and Physics 189
- Python Code Snippet 190
- Multiple Choice Questions 192

26 The Tolman-Oppenheimer-Volkoff Equation — 195
- Astrophysical Context 195
- Mathematical Derivation 195
- Analytical Solutions 196
- Numerical Methods 196
- Applications in Stellar Structure 196
- Python Code Snippet 197
- Multiple Choice Questions 199

27 The Yamabe Equation — 202
- Conformal Geometry Background 202
  1. Conformal Transformations 202
  2. Yamabe Flow 202
  3. Existence and Uniqueness 203
  4. Conformal Deformation and Positive Yamabe Constant 203
- Mathematical Formulation 203
  1. Yamabe Conformal Class 204
  2. Yamabe Problem 204
- Existence and Uniqueness 205
  1. Closed Manifolds with Constant Scalar Curvature 205
  2. General Manifolds and Boundary Conditions 205
- Numerical Methods 205

	1	Finite Difference Methods 206
	2	Variational Methods 206

Applications in Mathematical Physics 206
    1   Quantum Field Theory 206
    2   String Theory 207
    3   General Relativity 207
    4   Geometric Analysis 207
    5   Quantum Field Theory 207
    6   Functional Analysis 207
Conclusion 207
Python Code Snippet 208
Multiple Choice Questions 209

28 The Dirichlet Problem   213

Boundary Value Problem Setup 213
    1   Dirichlet Boundary Conditions 213
    2   The Dirichlet Problem 213
Weak Solutions and Variational Formulation 214
    1   Sobolev Spaces 214
    2   Weak Solutions 214
    3   Variational Formulation 215
Classical Solutions 215
    1   Finite Difference Methods 215
    2   Finite Element Methods 216
Numerical Methods 216
    1   Finite Difference Methods 216
    2   Finite Element Methods 216
Applications in Engineering 217
    1   Structural Analysis 217
    2   Heat Transfer 217
    3   Fluid Mechanics 217
    4   Electromagnetics 218
    5   Numerical Simulations 218
Conclusion 218
Python Code Snippet 219
Multiple Choice Questions 220

29 The Von Kármán Equations   224

Plate Theory in Structural Mechanics 224
    1   Assumptions and Governing Equations 224
    2   Boundary Conditions 225
    3   Significance in Structural Analysis 226

      Python Code Snippet 226
      Multiple Choice Questions 228

30 Maximum Principles for Elliptic Equations   231
      Statement of Maximum Principles 231
      1   The Strong Maximum Principle 231
      2   The Weak Maximum Principle 232
      Proofs and Mathematical Insights 232
      Applications to Uniqueness Theorems 233
      Discrete Maximum Principles 234
      Practical Applications 234
      Conclusion 235
      Python Code Snippet 236
      Multiple Choice Questions 238

31 Nonlinear Elliptic Equations   241
      Introduction and Examples 241
      1   Examples of Nonlinear Elliptic Equations . . 242
      Sobolev Spaces for Nonlinear Problems 243
      1   Definition of Sobolev Spaces 243
      2   Variational Formulations 243
      Existence Theorems 244
      Regularity and Stability 244
      1   Regularity and Stability of Solutions 245
      Applications in Nonlinear Analysis 245
      Conclusion 246
      Python Code Snippet 246
      Multiple Choice Questions 249

32 The De Giorgi-Nash-Moser Theorem   252
      Background and Importance 252
      Statement of the Theorem 253
      1   Monotonicity and Localization Properties . . 253
      2   Existence of Suitable Integrals 253
      3   Generalization to Nonlinear Systems and Higher
              Dimensions 254
      Regularity Theory and Applications 254
      Python Code Snippet 255
      Multiple Choice Questions 257

Chapter 1

Laplace's Equation

Historical Background and Importance

The study of elliptic partial differential equations has a long and rich history, dating back to the pioneering work of Laplace in the late 18th century. Laplace's equation, a specific type of elliptic equation, has been of great importance in various areas of science and engineering. Its significance lies in its ability to describe equilibrium and steady-state phenomena governed by Laplace's equation, such as electrostatics, heat conduction, and fluid flow.

Mathematical Formulation

Laplace's equation is a second-order linear partial differential equation of the form:
$$\Delta u = 0, \tag{1.1}$$
where Δ is the Laplacian operator defined as the divergence of the gradient, i.e.,
$$\Delta u = \nabla \cdot (\nabla u). \tag{1.2}$$
Here, u represents the unknown function to be solved. Laplace's equation can be defined in various dimensions and coordinate systems, such as in Cartesian, cylindrical, or spherical coordinates.

Fundamental Solutions

Finding the solutions to Laplace's equation requires understanding the fundamental solutions, which are solutions that satisfy Laplace's equation and possess certain boundary conditions. In three dimensions, the fundamental solution takes the form of the Green's function, denoted by $G(\mathbf{x}, \mathbf{x}')$, which satisfies:

$$\Delta G(\mathbf{x}, \mathbf{x}') = \delta(\mathbf{x} - \mathbf{x}'), \qquad (1.3)$$

where $\delta(\mathbf{x} - \mathbf{x}')$ is the Dirac delta function. The Green's function provides a means to solve Laplace's equation for arbitrary source terms via convolution.

Boundary Value Problems

Boundary value problems involving Laplace's equation arise when prescribing specific conditions on the boundaries of a domain. These conditions often take the form of Dirichlet, Neumann, or mixed boundary conditions. Dirichlet boundary conditions fix the value of the unknown function on the boundary, while Neumann conditions prescribe the normal derivative. Mixed boundary conditions consist of a combination of Dirichlet and Neumann conditions. Solving boundary value problems involves finding the appropriate combination of basic solutions and satisfying the given boundary conditions.

Applications in Electrostatics

One of the most important applications of Laplace's equation is in the field of electrostatics. In this context, Laplace's equation describes the electric potential in regions devoid of charge, known as electrostatics. By solving Laplace's equation subject to appropriate boundary conditions, one can determine the electric potential and subsequently obtain the electric field via differentiation. This has wide-ranging implications in various electrical systems, such as capacitors, conductors, and electrostatic shielding.

The study of Laplace's equation in electrostatics also includes the concept of harmonic functions, which are solutions to Laplace's equation that exhibit certain regularity properties. Harmonic functions play a fundamental role in the analysis and understanding of electrostatic phenomena, enabling the calculation of fields, potentials, and charges by exploiting the symmetries and properties of harmonic functions.

In summary, Laplace's equation forms the foundation for understanding a wide range of physical phenomena, particularly in the fields of electrostatics, heat conduction, and fluid flow. Its mathematical formulation and solutions provide valuable insights into equilibrium and steady-state problems. In the subsequent chapters, we will explore other elliptic partial differential equations and their practical applications."'latex

Python Code Snippet

Below is a Python code snippet that implements algorithms for solving Laplace's Equation using the Finite Difference Method. The code demonstrates how to create a grid for numerical computation, apply boundary conditions, and solve for the electrostatic potential.

```python
import numpy as np
import matplotlib.pyplot as plt

def initialize_grid(size, boundary_conditions):
    '''
    Initialize the grid with boundary conditions.
    :param size: Size of the grid as a tuple (rows, cols).
    :param boundary_conditions: Tuple containing boundary values
        (top, bottom, left, right).
    :return: Initialized grid with boundary conditions applied.
    '''
    grid = np.zeros(size)
    grid[0, :] = boundary_conditions[0]    # Top boundary
    grid[-1, :] = boundary_conditions[1]   # Bottom boundary
    grid[:, 0] = boundary_conditions[2]    # Left boundary
    grid[:, -1] = boundary_conditions[3]   # Right boundary
    return grid

def update_grid(grid, tolerance):
    '''
    Update the grid based on the finite difference method until
        convergence.
    :param grid: Current grid state.
    :param tolerance: Convergence criteria.
    :return: Updated grid after solving.
    '''
    rows, cols = grid.shape
    error = float('inf')
```

```python
    while error > tolerance:
        new_grid = grid.copy()  # Create a copy to store updates
        error = 0

        # Iterate over the interior points of the grid
        for i in range(1, rows - 1):
            for j in range(1, cols - 1):
                new_grid[i, j] = 0.25 * (grid[i + 1, j] + grid[i -
                ↪ 1, j] + grid[i, j + 1] + grid[i, j - 1])

        # Calculate the error
        error = np.max(np.abs(new_grid - grid))  # Maximum change
        grid = new_grid  # Update grid with new values

    return grid

def plot_potential(grid):
    '''
    Plot the electrostatic potential using matshow.
    :param grid: The grid containing potential values.
    '''
    plt.imshow(grid, cmap='hot', interpolation='nearest')
    plt.colorbar(label='Potential (V)')
    plt.title('Electric Potential Distribution')
    plt.xlabel('X-axis')
    plt.ylabel('Y-axis')
    plt.show()


# Parameters
grid_size = (50, 50)  # Size of the grid
boundary_conditions = (0, 0, 100, 0)  # Top, Bottom, Left, Right
↪ values
tolerance = 1e-4  # Tolerance for convergence

# Initialize grid and apply boundary conditions
potential_grid = initialize_grid(grid_size, boundary_conditions)

# Update grid using the finite difference method
final_potential = update_grid(potential_grid, tolerance)

# Plot the results
plot_potential(final_potential)
```

This code defines three functions:

- `initialize_grid` creates a grid and applies the specified boundary conditions to its edges.
- `update_grid` iteratively updates the grid until the potential values converge within a specified tolerance using the finite difference

method.
- `plot_potential` visualizes the computed electric potential across the grid.

The provided example initializes a grid for Laplace's equation with specified boundary conditions, iteratively solves for the electric potential, and then plots the resulting potential distribution.
""

Multiple Choice Questions

1. What type of equation is Laplace's equation?

    (a) Parabolic

    (b) Hyperbolic

    (c) Elliptic

    (d) Linear

2. Which of the following best describes the Laplacian operator?

    (a) The curl of a vector field

    (b) The divergence of the gradient of a scalar function

    (c) The integral of a function over a domain

    (d) The derivative of a function with respect to time

3. A function u that satisfies $\Delta u = 0$ is known as:

    (a) Harmonic

    (b) Continuous

    (c) Differentiable

    (d) Linear

4. When solving Laplace's equation, what type of boundary condition prescribes the values of a function on the boundary?

    (a) Neumann boundary condition

    (b) Dirichlet boundary condition

    (c) Mixed boundary condition

    (d) Natural boundary condition

5. In electrostatics, Laplace's equation is used to calculate which of the following?

(a) Charge distribution

(b) Electric field

(c) Electric potential

(d) Magnetic field

6. The Green's function for Laplace's equation is used for solving problems in:

    (a) Statistical mechanics

    (b) Quantum mechanics

    (c) Electrostatics

    (d) Linear algebra

7. Which of the following statements is true concerning harmonic functions?

    (a) They are always linear functions.

    (b) They can be defined only in two dimensions.

    (c) They minimize the average value over any region.

    (d) They do not have continuous derivatives.

Answers:

1. **C: Elliptic** Laplace's equation is classified as an elliptic partial differential equation, which is characterized by its solutions representing steady-state distributions.

2. **B: The divergence of the gradient of a scalar function** The Laplacian operator Δ is defined as the divergence of the gradient of a scalar function, giving insight into how the function's value spreads or changes in space.

3. **A: Harmonic** A function satisfying Laplace's equation $\Delta u = 0$ is known as a harmonic function, which has properties that are helpful in potential theory.

4. **B: Dirichlet boundary condition** The Dirichlet boundary condition specifies the values of a function on the boundary of the domain, commonly used in solving boundary value problems associated with Laplace's equation.

5. **C: Electric potential** In electrostatics, Laplace's equation is used to describe the electric potential in regions without free charges, and from the potential, the electric field can be derived through differentiation.

6. **C: Electrostatics** The Green's function for Laplace's equation is particularly utilized in electrostatics to solve boundary value problems, which involve determining potentials under given constraints.

7. **C: They minimize the average value over any region.** Harmonic functions are characterized by the mean value property, which states that their value at any point is equal to the average of their values over any surrounding surface. This property is crucial in multiple applications, including physics and engineering.

Chapter 2

Poisson's Equation

In this chapter, we delve into the study of Poisson's equation, a particular type of elliptic partial differential equation that arises in various scientific and engineering applications. Poisson's equation is closely related to Laplace's equation, but with an additional source term. It finds particular relevance in problems involving the distribution of scalar fields, such as potential, temperature, or concentration. Understanding the mathematical formulation, solution methods, and practical applications of Poisson's equation is essential for tackling a wide range of physical and mathematical problems.

Derivation and Physical Context

Poisson's equation is derived by introducing a source term into Laplace's equation. In its simplest form, it can be written as follows:

$$\Delta u = f(\mathbf{x}), \qquad (2.1)$$

where u is the unknown scalar function, Δ is the Laplacian operator, and $f(\mathbf{x})$ represents the given source function. Poisson's equation appears in various areas of physics and engineering, such as electrostatics, heat conduction, fluid dynamics, and potential theory.

Green's Functions

To solve Poisson's equation, we often employ the concept of Green's functions, which are fundamental solutions representing the influence of a point source. The Green's function, denoted as $G(\mathbf{x}, \mathbf{x}')$, satisfies the equation:

$$\Delta G(\mathbf{x}, \mathbf{x}') = \delta(\mathbf{x} - \mathbf{x}'), \tag{2.2}$$

where $\delta(\mathbf{x} - \mathbf{x}')$ is the Dirac delta function. The Green's function provides a powerful mathematical tool for solving Poisson's equation by convolving it with the source term.

Numerical Methods: Finite Difference

Numerical methods play a crucial role in solving Poisson's equation, especially when analytical solutions are infeasible or unavailable. One common numerical approach is the finite difference method, which approximates the derivatives in the equation using difference quotients. By discretizing the computational domain into a grid of points, we can derive algebraic equations that approximate Poisson's equation. Solving these equations yields the values of u at each grid point, providing an approximate solution to the problem.

Boundary Conditions

Solving Poisson's equation requires specifying suitable boundary conditions that determine the behavior of the scalar field on the domain's boundary. Dirichlet boundary conditions fix the value of u on the boundary, whereas Neumann boundary conditions prescribe the normal derivative of u. Mixed boundary conditions involve a combination of Dirichlet and Neumann conditions. Selecting appropriate boundary conditions is vital for obtaining unique solutions that satisfy the physical or mathematical requirements of the problem.

Applications in Heat Transfer

One prominent application of Poisson's equation lies in the study of heat transfer. By using Poisson's equation to model temperature distributions and heat conduction, we can analyze heat flow in various systems, such as conductive materials, cooling processes, and thermal insulation. Heat transfer problems often involve solving Poisson's equation with appropriate boundary conditions and thermal source terms. The resulting solutions provide insights into temperature distributions, heat fluxes, and thermal gradients within the system.

In summary, Poisson's equation plays a fundamental role in numerous scientific and engineering fields. By introducing a source term to Laplace's equation, Poisson's equation enables the modeling and analysis of scalar fields affected by external influences. The utilization of Green's functions, numerical methods, and suitable boundary conditions allows us to obtain solutions to Poisson's equation, providing valuable insights into the behavior of physical quantities in diverse applications, including heat transfer, fluid dynamics, electrostatics, and potential theory.Certainly! Below is a Python code snippet that implements the important equations and algorithms associated with Poisson's equation as discussed in the chapter.

Python Code Snippet

Below is a Python code snippet that implements the solution of Poisson's equation using the finite difference method along with a function to compute the Green's function for a 2D domain.

```python
import numpy as np
import matplotlib.pyplot as plt

def solve_poisson(f, boundary_conditions, grid_size, iterations):
    '''
    Solve Poisson's equation using the finite difference method.

    :param f: Function representing the source term (f).
    :param boundary_conditions: Values of the boundary (Dirichlet
        conditions).
    :param grid_size: Tuple representing the number of grid points
        along x and y directions.
    :param iterations: Number of iterations for the Jacobi method.
    :return: 2D array representing the solution u.
    '''
    # Create a grid
    u = np.zeros(grid_size)
    u[0, :] = boundary_conditions['top']      # Top boundary
    u[-1, :] = boundary_conditions['bottom']   # Bottom boundary
```

```python
        u[:, 0] = boundary_conditions['left']    # Left boundary
        u[:, -1] = boundary_conditions['right']  # Right boundary

        # Iterative Jacobi method
        for it in range(iterations):
            u_new = u.copy()
            for i in range(1, grid_size[0] - 1):
                for j in range(1, grid_size[1] - 1):
                    u_new[i, j] = 0.25 * (u[i + 1, j] + u[i - 1, j] +
                     ↪    u[i, j + 1] + u[i, j - 1] - f(i, j))
            u = u_new
        return u

def compute_green_function(x, y):
    '''
    Compute the Green's function for Poisson's equation in a 2D
     ↪   domain.

    :param x: x-coordinate.
    :param y: y-coordinate.
    :return: Value of the Green's function at (x, y).
    '''
    r = np.sqrt(x**2 + y**2)
    if r == 0:
        return 0  # Avoid singularity at the origin
    return -1 / (2 * np.pi * r)    # 2D Green's function for Poisson's
     ↪   equation


# Example parameters
grid_size = (50, 50)
iterations = 1000

# Source term function (f)
def source_term(i, j):
    # A point source in the center of the grid
    if (i == grid_size[0] // 2) and (j == grid_size[1] // 2):
        return 100  # Point source strength
    return 0

# Boundary conditions
boundary_conditions = {
    'top': 0,
    'bottom': 0,
    'left': 0,
    'right': 0
}

# Solve Poisson's equation
solution = solve_poisson(source_term, boundary_conditions,
 ↪   grid_size, iterations)

# Plotting the result
```

```
plt.imshow(solution, extent=(0, 1, 0, 1), origin='lower')
plt.colorbar(label='Potential u')
plt.title('Solution of Poisson\'s Equation')
plt.xlabel('x-axis')
plt.ylabel('y-axis')
plt.show()

# Compute and print the Green's function at a specific point
gx, gy = 0.1, 0.1  # Coordinates for Green's function
green_value = compute_green_function(gx, gy)
print("Value of Green's function at ({}, {}): {}".format(gx, gy,
    green_value))
```

This code defines two functions:

- `solve_poisson` utilizes the finite difference method to solve Poisson's equation over a specified grid, given a source term and boundary conditions.
- `compute_green_function` calculates the value of the Green's function at a specified point in a 2D domain.

The provided example uses a source term representing a point source in the center of the domain, applies Dirichlet boundary conditions, and visualizes the potential distribution resulting from this setup. Additionally, it computes the Green's function at a specified point and prints the result.

Multiple Choice Questions

1. Poisson's equation introduces which of the following components into Laplace's equation?

    (a) A time-dependent term

    (b) A boundary term

    (c) A source term

    (d) A damping term

2. What is the general form of Poisson's equation?

    (a) $\Delta u = 0$

    (b) $\Delta u = f(\mathbf{x})$

    (c) $\Delta u + ku = 0$

    (d) $\Delta u = -\nabla^2 u$

3. Which method is commonly used to derive numerical solutions for Poisson's equation?

    (a) Finite Element Method
    (b) Calculus of Variations
    (c) Green's Theorem
    (d) Fourier Transform

4. In the context of solving Poisson's equation, what role do Green's functions play?

    (a) They represent approximations of linear mappings.
    (b) They serve as fundamental solutions to relate point sources to potential fields.
    (c) They are used for evaluating integrals over arbitrary domains.
    (d) They simplify boundary conditions to Dirichlet types.

5. Which of the following is NOT a type of boundary condition relevant for Poisson's equation?

    (a) Dirichlet boundary condition
    (b) Neumann boundary condition
    (c) Robin boundary condition
    (d) Mixed boundary condition

6. Poisson's equation is commonly applied in which field?

    (a) Quantum mechanics
    (b) Structural analysis
    (c) Heat transfer
    (d) Cryptography

7. The significance of the source term $f(\mathbf{x})$ in Poisson's equation is that it:

    (a) Represents external influences affecting the scalar field.
    (b) Is always treated as a constant.
    (c) Is relevant only for boundary value problems.
    (d) Has no effect on the solution of the equation.

Answers:

1. **C: A source term** Poisson's equation includes an additional source term $f(\mathbf{x})$ compared to Laplace's equation, allowing it to account for influences like charge density in electrostatics or heat sources in temperature distributions.

2. **B: $\Delta u = f(\mathbf{x})$** The general form of Poisson's equation states that the Laplacian of u corresponds to a source term $f(\mathbf{x})$, capturing the essence of how sources influence the scalar field.

3. **A: Finite Element Method** The Finite Element Method is widely used to derive numerical solutions for Poisson's equation, particularly when dealing with complex geometries and boundary conditions.

4. **B: They serve as fundamental solutions to relate point sources to potential fields.** Green's functions act as a bridge that relates the response of the system (the scalar potential) to point sources represented by the Dirac delta function, facilitating the solution of Poisson's equation.

5. **D: Mixed boundary condition** While Dirichlet, Neumann, and Robin conditions are common, "mixed boundary condition" typically refers to a combination of conditions, whereas the question intends to find a type that does not fit traditional classifications.

6. **C: Heat transfer** Poisson's equation is often applied in heat transfer problems, where it models the distribution of temperature in materials influenced by heat sources.

7. **A: Represents external influences affecting the scalar field.** The source term $f(\mathbf{x})$ in Poisson's equation represents external influences, such as volumetric charge density in electrostatics or heat sources in thermal analysis, which directly affect the distribution of the scalar quantity u.

Chapter 3

The Helmholtz Equation

The Helmholtz equation, named after the German physicist Hermann von Helmholtz, is a fundamental partial differential equation that appears in various areas of physics, particularly in wave propagation problems. It is an elliptic equation that combines elements of both the Laplace and wave equations. In this chapter, we will explore the mathematical setup, solution techniques, and physical significance of the Helmholtz equation, as well as its applications in acoustics, electromagnetics, and other fields.

Mathematical Setup and Physical Significance

The Helmholtz equation in its general form is given by:

$$\Delta u + k^2 u = 0, \tag{3.1}$$

where u is the unknown scalar function, Δ represents the Laplacian operator, and k is the wave number. The Helmholtz equation arises in numerous physical phenomena involving the propagation of waves or oscillations, such as acoustics, optics, electromagnetics, and quantum mechanics.

The wave number k is related to the wavelength λ by the equation:

$$\lambda = \frac{2\pi}{k}. \tag{3.2}$$

A smaller value of k corresponds to a longer wavelength, while a larger value of k represents a shorter wavelength. The wavelength determines the spatial periodicity of the wave or oscillation.

Separation of Variables

The separation of variables technique is a powerful method widely used to solve the Helmholtz equation. It takes advantage of the assumed separability of the solution $u(\mathbf{x})$ into spatial components, usually expressed as:

$$u(\mathbf{x}) = X(x)Y(y)Z(z), \qquad (3.3)$$

where x, y, and z represent the Cartesian coordinates in three dimensions. Substituting this assumed form into the Helmholtz equation (3.1), we obtain three separate ordinary differential equations (ODEs) for the functions X, Y, and Z, respectively. Solving these ODEs yields a general solution for $u(\mathbf{x})$ that satisfies the Helmholtz equation.

Methods of Fundamental Solutions

In addition to the separation of variables method, the Helmholtz equation can also be solved using the Method of Fundamental Solutions (MFS). This powerful numerical technique is based on representing the solution as a linear combination of fundamental solutions centered at certain points in the domain. The MFS provides an efficient and accurate approach to solving the Helmholtz equation, particularly in problems with complex geometries or irregular domains.

Eigenvalue Problems

Solving the Helmholtz equation often involves seeking solutions for which certain specific values of k (known as eigenvalues) yield nontrivial solutions. This gives rise to eigenvalue problems associated with the Helmholtz equation. The determination of eigenvalues and corresponding eigenfunctions is crucial for understanding the fundamental modes of wave propagation and for characterizing the behavior of physical systems governed by the Helmholtz equation.

Applications in Acoustics

One of the primary applications of the Helmholtz equation is in the field of acoustics, which deals with the study of sound and its propagation. By modeling sound waves using the Helmholtz equation, we can analyze the behavior of acoustic fields in various configurations. This includes problems related to the scattering of sound by obstacles, the resonance in cavities, the study of sound absorption and transmission, and more generally, the characterization of sound propagation in heterogeneous media.

In summary, the Helmholtz equation provides a mathematical framework for understanding wave propagation phenomena and oscillatory behavior in various physical systems. By applying separation of variables, the method of fundamental solutions, or other solution techniques, we can obtain solutions that describe the behavior of waves or oscillations in different domains. The Helmholtz equation finds wide-ranging applications in acoustics, electromagnetics, optics, quantum mechanics, and other areas, allowing us to explore and analyze various physical phenomena in depth.Here is a comprehensive Python code snippet that implements important equations, algorithms, and formulas mentioned in the chapter on the Helmholtz equation. This includes a function to solve the Helmholtz equation using separation of variables and methods for calculating the eigenvalues and eigenfunctions.

Python Code Snippet

Below is a Python code snippet that calculates the solutions for the Helmholtz equation using separation of variables as well as methods to compute eigenvalues and eigenfunctions in a rectangular domain.

```python
import numpy as np
import matplotlib.pyplot as plt

def helmholtz_separation_of_variables(x, y, k):
    '''
    Compute the solution of the Helmholtz equation using separation
        of variables.
    :param x: 1D array of x-coordinates.
    :param y: 1D array of y-coordinates.
    :param k: Wave number.
    :return: 2D array representing the solution u(x,y).
    '''
```

```python
    X = np.sin(k * x)    # Separation of variables in x
    Y = np.sin(k * y)    # Separation of variables in y
    u = np.outer(X, Y)   # Outer product to get u(x,y)
    return u

def compute_eigenvalues_and_eigenfunctions(n, m):
    '''
    Calculate eigenvalues and eigenfunctions for a rectangular
    ↪ domain.
    :param n: Number of eigenfunctions in the x-direction.
    :param m: Number of eigenfunctions in the y-direction.
    :return: List of eigenvalues and associated eigenfunctions.
    '''
    eigenvalues = []
    eigenfunctions = []

    for i in range(1, n + 1):
        for j in range(1, m + 1):
            k = np.sqrt(i**2 + j**2)   # Calculate the eigenvalue
            eigenvalues.append(k)
            eigenfunctions.append((i, j))   # Store the mode indices

    return eigenvalues, eigenfunctions

def plot_solution(x, y, u):
    '''
    Plot the solution of the Helmholtz equation.
    :param x: 1D array of x-coordinates.
    :param y: 1D array of y-coordinates.
    :param u: 2D array of the solution u(x,y).
    '''
    plt.figure(figsize=(8, 6))
    plt.contourf(x, y, u, cmap='viridis')
    plt.colorbar(label='u(x,y)')
    plt.title('Solution of the Helmholtz Equation')
    plt.xlabel('x')
    plt.ylabel('y')
    plt.show()

# Parameters for the calculations
k = 2 * np.pi / 1   # Wave number for wavelength = 1
x = np.linspace(0, 3, 100)   # x-coordinates
y = np.linspace(0, 3, 100)   # y-coordinates

# Calculate the solution of the Helmholtz equation
u = helmholtz_separation_of_variables(x, y, k)

# Compute eigenvalues and eigenfunctions
n = 5   # Number of eigenfunctions in the x-direction
m = 5   # Number of eigenfunctions in the y-direction
eigenvalues, eigenfunctions =
↪   compute_eigenvalues_and_eigenfunctions(n, m)
```

```
# Plot the results
plot_solution(x, y, u)

# Output results
print("Eigenvalues:", eigenvalues)
print("Eigenfunctions (mode indices):", eigenfunctions)
```

This code defines several functions:

- helmholtz_separation_of_variables computes the solution of the Helmholtz equation using separation of variables in a 2D rectangular domain.
- compute_eigenvalues_and_eigenfunctions calculates the eigenvalues and eigenfunctions for a rectangular domain based on the specified number of modes in both dimensions.
- plot_solution visualizes the resulting solution $u(x, y)$ of the Helmholtz equation.

The provided code calculates the solution to the Helmholtz equation and the corresponding eigenvalues and eigenfunctions for a specified rectangle and then displays the solution as a contour plot while printing the eigenvalues and mode indices.

Multiple Choice Questions

1. What is the general form of the Helmholtz equation?

    (a) $\Delta u - k^2 u = 0$
    (b) $\Delta u + k^2 u = 0$
    (c) $\Delta u + k^2 u = f(x)$
    (d) $\Delta u - k^2 u = f(x)$

2. The wave number k is related to the wavelength λ by which equation?

    (a) $\lambda = \frac{k}{2\pi}$
    (b) $\lambda = \frac{2\pi}{k}$
    (c) $\lambda = k^2$
    (d) $\lambda = k - 2\pi$

3. In the method of separation of variables, the solution $u(\mathbf{x})$ is assumed to be of the form:

(a) $u(\mathbf{x}) = X(x, y, z)$

(b) $u(\mathbf{x}) = X(x)Y(y)Z(z)$

(c) $u(\mathbf{x}) = Ae^{ikx}$

(d) $u(\mathbf{x}) = A\sin(kx)$

4. Eigenvalue problems associated with the Helmholtz equation primarily focus on:

   (a) Finding trivial solutions.

   (b) Determining specific values of k that yield nontrivial solutions.

   (c) Determining the spatial domains for solutions.

   (d) Computing the Laplacian of eigenfunctions.

5. Which of the following is NOT a common application of the Helmholtz equation?

   (a) Acoustics

   (b) Electromagnetics

   (c) Heat conduction

   (d) Quantum mechanics

6. The method of fundamental solutions (MFS) is particularly valuable for:

   (a) Solving only linear PDEs.

   (b) Geographic information systems (GIS).

   (c) Problems with complex geometries or irregular domains.

   (d) Simple geometries with no boundaries.

7. Which of the following statements about wave number k is true?

   (a) A larger k value corresponds to a longer wavelength.

   (b) k is always a positive integer.

   (c) k represents the spatial frequency of the wave.

   (d) k is irrelevant in the context of wave propagation.

Answers:

1. **B:** $\Delta u + k^2 u = 0$ The Helmholtz equation is correctly defined as $\Delta u + k^2 u = 0$, indicating the relationship between the Laplacian of a function and the wave number, with a negative sign in front of the $k^2 u$ term.

2. **B:** $\lambda = \frac{2\pi}{k}$ This formula represents the relationship between wavelength λ and wave number k. As the wave number increases, the wavelength decreases.

3. **B:** $u(\mathbf{x}) = X(x)Y(y)Z(z)$ The method of separation of variables decomposes $u(\mathbf{x})$ into independent spatial functions X, Y, and Z, allowing for the solution of each part individually.

4. **B: Determining specific values of k that yield nontrivial solutions** Eigenvalue problems seek specific values of k (eigenvalues) for which there are nontrivial solutions to the Helmholtz equation, revealing critical modes of wave propagation.

5. **C: Heat conduction** The Helmholtz equation is primarily related to wave phenomena, whereas heat conduction is described by the diffusion equation, making this the correct response.

6. **C: Problems with complex geometries or irregular domains** The method of fundamental solutions is particularly advantageous for handling problems with complex boundary conditions and geometries, making it a powerful computational technique.

7. **C: k represents the spatial frequency of the wave** The wave number k indicates how many wavelengths fit into a given spatial dimension, effectively representing the spatial frequency. A higher k means more cycles per unit distance, hence a shorter wavelength.

Chapter 4

The Biharmonic Equation

The Biharmonic equation, also known as the biharmonic Dirichlet problem, is a fundamental partial differential equation that describes various phenomena in elasticity theory. The equation is a fourth-order elliptic partial differential equation that involves the Laplacian operator applied twice. In this chapter, we will explore the mathematical formulation, boundary conditions, and solution techniques for the biharmonic equation. We will also discuss its applications in elasticity theory and its numerical implementation.

Governing Equations in 2D and 3D

The biharmonic equation in two dimensions is given by:

$$\Delta^2 u = \frac{\partial^4 u}{\partial x^4} + 2\frac{\partial^4 u}{\partial x^2 \partial y^2} + \frac{\partial^4 u}{\partial y^4} = 0, \qquad (4.1)$$

where $u(x, y)$ is the unknown scalar function, and Δ denotes the Laplacian operator.

In three dimensions, the biharmonic equation is expressed as:

$$\Delta^2 u = \frac{\partial^4 u}{\partial x^4} + 2\frac{\partial^4 u}{\partial x^2 \partial y^2} + \frac{\partial^4 u}{\partial y^4} + 2\frac{\partial^4 u}{\partial x^2 \partial z^2} + 2\frac{\partial^4 u}{\partial y^2 \partial z^2} + \frac{\partial^4 u}{\partial z^4} = 0. \qquad (4.2)$$

The biharmonic equation governs various phenomena involving elasticity, such as the bending of thin beams or plates, the buckling of structures, and the deformation of elastic bodies.

Boundary Conditions and Their Types

The biharmonic equation is a boundary value problem, which means that it requires additional conditions to determine a unique solution. The most common boundary condition is the Dirichlet boundary condition, which specifies the value of the function on the boundary of the domain. For example, in a rectangular domain Ω defined by $a \leq x \leq b$ and $c \leq y \leq d$, the Dirichlet boundary condition is given as:

$$u(x,y) = g(x,y) \quad \text{for } (x,y) \in \partial\Omega, \tag{4.3}$$

where $g(x,y)$ is a known function defined on the boundary $\partial\Omega$.

Other types of boundary conditions for the biharmonic equation include the Neumann boundary condition, which specifies the derivative of the function on the boundary, and the mixed boundary condition, which combines both Dirichlet and Neumann conditions.

Green's Function Approach

The biharmonic equation can be solved using the Green's function approach. Green's functions are fundamental solutions that satisfy the biharmonic equation and certain boundary conditions. By convolving the Green's function with the source term, we can obtain the solution to the biharmonic equation.

In two dimensions, the Green's function $G(x,y;\xi,\eta)$ for the biharmonic equation is given by:

$$\begin{aligned} G(x,y;\xi,\eta) = &- (2\pi)^{-2} \left(|x-\xi|^2 - (x-\xi)^2(x-\xi) \right. \\ &\left. + |y-\eta|^2 - (y-\eta)^2(y-\eta) \right). \end{aligned} \tag{4.4}$$

In three dimensions, the Green's function $G(x,y,z;\xi,\eta,\zeta)$ is more complicated and involves additional terms.

Using the Green's function, the solution $u(x,y)$ to the biharmonic equation with Dirichlet boundary conditions can be expressed as:

$$u(x,y) = -\iint_\Omega G(x,y;\xi,\eta) f(\xi,\eta) d\xi d\eta + \int_{\partial\Omega} \left(\frac{\partial G}{\partial n} - G\frac{\partial}{\partial n} \right) g(\xi,\eta) dS, \tag{4.5}$$

where $f(x,y)$ is the source term and $\frac{\partial}{\partial n}$ represents the derivative in the outward normal direction.

Numerical Implementation

Numerical methods are often employed to approximate the solution of the biharmonic equation, especially for complex geometries or domains where analytical solutions are not available. Finite difference methods, finite element methods, and spectral methods are commonly used for the numerical implementation of the biharmonic equation.

Finite difference methods discretize the domain into a grid and approximate the derivatives using finite difference approximations. The resulting system of algebraic equations can be solved using various techniques.

Finite element methods, on the other hand, divide the domain into smaller elements and use variational principles to obtain local equations. These equations are then assembled to form a global system that can be solved for the unknowns.

Spectral methods utilize high-degree polynomials or other orthogonal functions as basis functions to approximate the solution. By applying appropriate interpolation techniques, the solution is represented in terms of these basis functions, and the equations are solved using linear algebra methods.

Applications in Elasticity Theory

The biharmonic equation finds significant applications in elasticity theory, particularly in the analysis of bending and deformation of elastic structures. By solving the biharmonic equation, we can study the behavior of thin beams, plates, or shells under various loading conditions. The biharmonic equation helps to determine the shape, deflection, and stress distribution of these structures, which are essential for designing safe and efficient engineering systems.

Additionally, the biharmonic equation is used in the study of buckling phenomena, which occur when a structure suddenly collapses or deforms under compression or bending. Understanding buckling is crucial for designing structures that can withstand loadings without failure.

1 Example Application: Bending of a Thin Beam

To illustrate the application of the biharmonic equation, let's consider the bending of a thin beam under a distributed load. The biharmonic equation can be used to determine the deflection of the beam and the corresponding stress distribution. Given a beam with length L and width W, subjected to a distributed load $q(x)$ along its length, the biharmonic equation for the deflection $w(x, y)$ can be written as:

$$\frac{\partial^4 w}{\partial x^4} + 2\frac{\partial^4 w}{\partial x^2 \partial y^2} + \frac{\partial^4 w}{\partial y^4} = \frac{q(x)}{D}, \qquad (4.6)$$

where D is the flexural rigidity of the beam, which depends on the material properties.

By solving Equation (4.6) with appropriate boundary conditions, we can determine the deflection $w(x, y)$ and subsequently calculate the stress distribution in the beam.

Conclusion

In this chapter, we explored the biharmonic equation, a fourth-order elliptic partial differential equation that arises in elasticity theory. We discussed the governing equations in 2D and 3D, various types of boundary conditions, and solution techniques for the biharmonic equation. Additionally, we highlighted the Green's function approach and numerical methods commonly employed for the numerical implementation of the biharmonic equation. Finally, we discussed the significant applications of the biharmonic equation in elasticity theory, focusing on the bending of thin beams as an example application.

Python Code Snippet

Below is a Python code snippet that implements the solution for the biharmonic equation using the finite difference method. This implementation includes the governing equations, boundary conditions, and the numerical solver.

```
import numpy as np
import matplotlib.pyplot as plt
```

```python
def biharmonic_solver(Lx, Ly, nx, ny, q_func, bc_func, D):
    '''
    Solves the biharmonic equation using the finite difference
    ↪ method.

    :param Lx: Length of the beam in the x direction.
    :param Ly: Width of the beam in the y direction.
    :param nx: Number of grid points in the x direction.
    :param ny: Number of grid points in the y direction.
    :param q_func: Function representing the distributed load.
    :param bc_func: Function for boundary conditions.
    :param D: Flexural rigidity of the beam.
    :return: The deflection u of the beam.
    '''
    # Create grid
    dx = Lx / (nx - 1)
    dy = Ly / (ny - 1)
    u = np.zeros((ny, nx))

    # Apply boundary conditions
    u[0, :] = bc_func(0, np.linspace(0, Lx, nx))    # Top boundary
    u[-1, :] = bc_func(Ly, np.linspace(0, Lx, nx))  # Bottom
    ↪ boundary
    u[:, 0] = bc_func(np.linspace(0, Ly, ny), 0)    # Left boundary
    u[:, -1] = bc_func(np.linspace(0, Ly, ny), Lx)  # Right
    ↪ boundary

    # Fill the interior grid with finite difference method
    for it in range(10000):  # Iteration for relaxation
        u_prev = u.copy()
        for j in range(1, ny - 1):
            for i in range(1, nx - 1):
                laplacian = (u_prev[j+1, i] - 2*u_prev[j, i] +
                ↪ u_prev[j-1, i]) / dy**2 + \
                            (u_prev[j, i+1] - 2*u_prev[j, i] +
                            ↪ u_prev[j, i-1]) / dx**2
                u[j, i] = u_prev[j, i] + (q_func(i*dx, j*dy) / (D))
                ↪ * (dx**2 * dy**2) / 2 - laplacian
        u[0, :] = bc_func(0, np.linspace(0, Lx, nx))  # Reapply
        ↪ boundary conditions
        u[-1, :] = bc_func(Ly, np.linspace(0, Lx, nx))
        u[:, 0] = bc_func(np.linspace(0, Ly, ny), 0)
        u[:, -1] = bc_func(np.linspace(0, Ly, ny), Lx)

    return u

def distributed_load(x, y):
    ''' Example distributed load function. '''
    return -10.0  # Constant load

def boundary_conditions(x, y):
    ''' Example boundary condition function. '''
    return 0.0  # Simply supported boundary
```

```
# Parameters for the domain
Lx = 1.0    # Length of the beam
Ly = 0.2    # Width of the beam
nx = 20     # Number of points in x
ny = 4      # Number of points in y
D = 1.0     # Flexural rigidity of the beam

# Solve the biharmonic equation
deflection = biharmonic_solver(Lx, Ly, nx, ny, distributed_load,
 ↪ boundary_conditions, D)

# Create a meshgrid for plotting
x = np.linspace(0, Lx, nx)
y = np.linspace(0, Ly, ny)
X, Y = np.meshgrid(x, y)

# Plotting the deflection
plt.figure(figsize=(10,6))
cp = plt.contourf(X, Y, deflection, levels=50, cmap='viridis')
plt.colorbar(cp)
plt.title('Deflection of Beam under Distributed Load')
plt.xlabel('Length (x)')
plt.ylabel('Width (y)')
plt.show()
```

This code defines the following functions:

- `biharmonic_solver` implements the finite difference method to solve the biharmonic equation given the length, width, grid points, distributed load, and boundary conditions.
- `distributed_load` specifies the distributed load acting on the beam.
- `boundary_conditions` defines the boundary conditions of the beam.

The provided example calculates the deflection of a simply supported beam under a constant distributed load and visualizes the results using a contour plot.

Multiple Choice Questions

1. Which equation represents the biharmonic equation in two dimensions?

   (a) $\Delta^2 u = \nabla^4 u = 0$

   (b) $\Delta^2 u = \frac{\partial^4 u}{\partial x^4} + 2\frac{\partial^4 u}{\partial x^2 \partial y^2} + \frac{\partial^4 u}{\partial y^4} = 0$

(c) $\Delta^2 u = \frac{\partial^2 u}{\partial x^2} + \frac{\partial^2 u}{\partial y^2} = 0$

(d) $\Delta^2 u = 0$

2. What boundary condition specifies the value of the function on the boundary of the domain?

    (a) Neumann boundary condition
    (b) Dirichlet boundary condition
    (c) Mixed boundary condition
    (d) Robin boundary condition

3. The Green's function for the biharmonic equation serves what purpose?

    (a) It provides a numerical approximation of the solution.
    (b) It defines the boundary of the domain.
    (c) It represents the fundamental solution to the equation.
    (d) It is used to compute eigenfunctions.

4. For the bending of a thin beam, the biharmonic equation relates the deflection $w(x, y)$ to which physical quantity?

    (a) Material density
    (b) Distributed load $q(x)$
    (c) Shear force
    (d) Internal energy

5. Which of the following numerical methods is commonly used to solve the biharmonic equation?

    (a) Finite difference method
    (b) Perturbation method
    (c) Shooting method
    (d) Random walk method

6. The biharmonic equation is commonly applied in which of the following fields?

    (a) Electromagnetics
    (b) Fluid dynamics
    (c) Elasticity theory

(d) Population dynamics

7. In elasticity theory, the biharmonic equation is primarily used to analyze:

    (a) Thermal diffusion in solid materials
    (b) Bending and deformation of elastic structures
    (c) Flow of fluids in porous media
    (d) Electromagnetic field distributions

Answers:

1. **B:** $\Delta^2 u = \frac{\partial^4 u}{\partial x^4} + 2\frac{\partial^4 u}{\partial x^2 \partial y^2} + \frac{\partial^4 u}{\partial y^4} = 0$ This option is the correct mathematical representation of the biharmonic equation in two dimensions.

2. **B: Dirichlet boundary condition** The Dirichlet boundary condition specifically requires the value of the function u to be given on the boundary of the domain, thus uniquely determining the solution.

3. **C: It represents the fundamental solution to the equation.** Green's functions describe how to construct solutions to differential equations and are powerful tools for solving boundary value problems, including the biharmonic equation.

4. **B: Distributed load** $q(x)$ The biharmonic equation relates the deflection of the beam to the applied distributed load, which influences its bending and stress distribution.

5. **A: Finite difference method** The finite difference method is a common numerical approach to approximate solutions to partial differential equations, including the biharmonic equation.

6. **C: Elasticity theory** The biharmonic equation is primarily encountered in elasticity theory, particularly in analyzing materials under stress or strain, such as beams and plates.

7. **B: Bending and deformation of elastic structures** In elasticity theory, the biharmonic equation addresses the deflection and deformation behavior of structures when subjected to forces, making it crucial for engineering applications.

Chapter 5

The Diffusion Equation

The diffusion equation is a fundamental partial differential equation that describes the behavior of a variety of diffusive processes, such as heat conduction, mass transfer, and molecular diffusion. In this chapter, we will delve into the mathematical derivation of the diffusion equation, explore its analytical solutions, and discuss numerical methods for solving it. Additionally, we will examine the stability and convergence properties of these numerical methods and illustrate the applications of the diffusion equation in biological processes.

Mathematical Derivation

The diffusion equation can be derived using fundamental principles of conservation. Consider a substance with concentration $u(x,t)$ that diffuses through a medium in one spatial dimension. The diffusion process can be modeled by considering the change in concentration with respect to both space and time. Based on Fick's law, the flux of the substance is proportional to the concentration gradient:

$$J(x,t) = -D\frac{\partial u(x,t)}{\partial x}, \qquad (5.1)$$

where $J(x,t)$ represents the mass flow per unit area, and D is the diffusion coefficient.

Applying the principle of conservation of mass, we can express the change in concentration within an infinitesimal volume element as the difference between the inflow and outflow of the substance:

$$\frac{\partial u(x,t)}{\partial t} = -\frac{\partial J(x,t)}{\partial x}. \tag{5.2}$$

Substituting Equation (5.1) into Equation (5.2), we obtain the diffusion equation:

$$\frac{\partial u(x,t)}{\partial t} = D\frac{\partial^2 u(x,t)}{\partial x^2}. \tag{5.3}$$

The diffusion equation governs the spatiotemporal evolution of the concentration distribution $u(x,t)$ in various diffusive processes.

Analytical Solutions

The diffusion equation has several exact analytical solutions that depend on the boundary and initial conditions. One of the simplest solutions corresponds to the case of a constant initial concentration $u(x,0) = u_0$ in an infinite domain with no flux at the boundaries, known as the "heat equation." In one spatial dimension, the heat equation is given by:

$$\frac{\partial u(x,t)}{\partial t} = \alpha \frac{\partial^2 u(x,t)}{\partial x^2}, \tag{5.4}$$

where $\alpha = \frac{D}{\rho c}$ represents the thermal diffusivity, with ρ as the material density and c as the specific heat.

The solution to the heat equation with an initial condition $u(x,0) = u_0$ and no-flux boundary conditions can be obtained using separation of variables and Fourier series expansion methods, leading to the well-known solution in one dimension:

$$u(x,t) = \sum_{n=1}^{\infty} A_n \exp\left(-\alpha \left(\frac{n\pi}{L}\right)^2 t\right) \sin\left(\frac{n\pi x}{L}\right), \tag{5.5}$$

where A_n are constants determined by the initial condition, L is the spatial domain length, and n represents the mode number.

Other analytical solutions to the diffusion equation include the Gaussian solution, the error function solution, and solutions in higher dimensions. These solutions provide insights into the behavior of diffusion processes under specific conditions.

Numerical Methods: Finite Element

Numerical methods are often employed to approximate the solution to the diffusion equation when analytical solutions are not available or feasible. One widely used method is the finite element method, which discretizes the domain into smaller finite elements and represents the concentration as a piecewise continuous function in terms of interpolation functions or shape functions.

By substituting the approximated solution into the diffusion equation and applying appropriate boundary and initial conditions, a system of algebraic equations is obtained. This system is then solved by matrix methods, such as Gaussian elimination or iterative solvers, to determine the concentration values at the nodes.

The finite element method offers flexibility in handling complex geometries and boundary conditions and provides accurate solutions for diffusion problems with varying coefficients or irregular domains. It has found extensive applications in diverse fields, including heat transfer, fluid dynamics, and biomedical engineering.

Stability and Convergence

In the numerical approximation of the diffusion equation, stability and convergence are crucial properties to ensure accurate and reliable results. Stability ensures that small errors do not lead to significant growth over time, while convergence guarantees that the numerical solution approaches the true solution as the discretization becomes finer.

For the finite element method, stability and convergence can be analyzed using mathematical techniques, such as energy estimates or the Lax-Richtmyer equivalence theorem. Additionally, certain numerical parameters, such as the time step and element size, must satisfy stability conditions, such as the CFL condition, to ensure accurate and stable solutions.

Numerical convergence can be assessed by examining the error between the numerical and analytical solutions, typically measured in terms of the L^2 norm or the maximum norm. Convergence studies involving grid refinement analysis are essential to validate the accuracy and convergence rate of the numerical method.

Applications in Biological Processes

The diffusion equation finds extensive applications in various biological processes, where the transfer of substances through cells, tissues, or organs significantly affects their function. Examples of biological systems where the diffusion equation is relevant include oxygen and nutrient transport in living tissues, drug delivery and distribution in the body, and cell migration dynamics.

Understanding the spatiotemporal distribution of concentrations and the transport mechanisms is crucial for developing effective therapies, predicting biological responses, and optimizing drug delivery systems. The diffusion equation, coupled with appropriate biological models and experimental data, enables the quantitative analysis and prediction of diffusion-driven processes in biological systems.

Summary

In this chapter, we explored the diffusion equation, a fundamental partial differential equation that governs diffusive processes in various fields. We derived the equation from fundamental conservation principles, discussed exact analytical solutions, and examined the finite element method as a numerical approach to solving the diffusion equation. Moreover, we emphasized the importance of stability and convergence in numerical methods and highlighted the applications of the diffusion equation in biological processes.

Python Code Snippet

Below is a Python code snippet that implements the important equations and algorithms discussed in this chapter regarding the diffusion equation, including the derivation, analytical solution (Gaussian solution), and finite element method for numerical approximation.

"'latex

```
import numpy as np
import matplotlib.pyplot as plt

def diffusion_analytical_solution(x, t, D, u0):
    '''
```

```python
    Calculate the analytical solution of the diffusion equation.
    :param x: Spatial position (1D numpy array).
    :param t: Time (scalar).
    :param D: Diffusion coefficient (scalar).
    :param u0: Initial concentration (scalar).
    :return: Concentration at position x and time t (numpy array).
    '''
    return u0 * np.exp(-x**2 / (4 * D * t)) / np.sqrt(4 * np.pi * D
    ↪    * t)

def diffusion_finite_element_method(nx, nt, L, T, D, u_init):
    '''
    Solve the diffusion equation using the finite element method.
    :param nx: Number of spatial grid points.
    :param nt: Number of time steps.
    :param L: Length of the spatial domain.
    :param T: Total time.
    :param D: Diffusion coefficient.
    :param u_init: Initial concentration distribution (1D numpy
    ↪    array).
    :return: Concentration history (2D numpy array: time x space).
    '''
    dx = L / (nx - 1)
    dt = T / nt
    r = D * dt / dx**2

    u = np.zeros((nt, nx))
    u[0, :] = u_init

    for n in range(0, nt - 1):
        for i in range(1, nx - 1):
            u[n + 1, i] = u[n, i] + r * (u[n, i + 1] - 2 * u[n, i] +
            ↪    u[n, i - 1])

    return u

# Parameters
D = 1.0    # Diffusion coefficient
L = 10.0   # Length of the domain
T = 1.0    # Total time
nx = 100   # Number of spatial points
nt = 1000  # Number of time steps
u0 = 1.0   # Initial concentration

# Initial concentration distribution
x = np.linspace(0, L, nx)
u_init = np.zeros(nx)
u_init[int(nx / 2)] = u0   # Point source in the middle

# Calculate the analytical solution
t_values = [0.01, 0.1, 0.5, 1.0]
for t in t_values:
    u_a = diffusion_analytical_solution(x, t, D, u0)
```

```
    plt.plot(x, u_a, label=f'Analytical at t={t}')

# Calculate the numerical solution using finite element method
u_numerical = diffusion_finite_element_method(nx, nt, L, T, D,
↪ u_init)

# Plot numerical results at final time step
plt.plot(x, u_numerical[-1, :], 'o-', label='Numerical at t=T',
↪ markersize=3)

# Plot settings
plt.title("Diffusion Equation Solutions")
plt.xlabel("Position (x)")
plt.ylabel("Concentration (u)")
plt.legend()
plt.grid()
plt.show()
```

"'

This code defines two primary functions:

- `diffusion_analytical_solution` calculates the analytical solution of the diffusion equation using the Gaussian function.
- `diffusion_finite_element_method` implements the finite element method to numerically solve the diffusion equation given initial and boundary conditions.

The provided example sets parameters for a diffusion problem, computes both analytical and numerical solutions, and then visualizes the results over time, demonstrating the behavior of the concentration as it diffuses through the medium.

Multiple Choice Questions

1. The diffusion equation can be derived from which fundamental principle?

    (a) Conservation of momentum

    (b) Conservation of energy

    (c) Conservation of mass

    (d) Conservation of charge

2. What is the form of the one-dimensional diffusion equation?

    (a) $\frac{\partial u(x,t)}{\partial t} = D \frac{\partial^2 u(x,t)}{\partial x^2}$

(b) $\frac{\partial u(x,t)}{\partial t} = D\frac{\partial u(x,t)}{\partial x}$

(c) $\frac{\partial^2 u(x,t)}{\partial t^2} = D\frac{\partial^2 u(x,t)}{\partial x^2}$

(d) $\frac{\partial u(x,t)}{\partial t} = u(x,t)\frac{\partial u(x,t)}{\partial x}$

3. Which of the following represents Fick's law?

   (a) $J = D\frac{\partial u}{\partial x}$

   (b) $J = -D\frac{\partial u}{\partial x}$

   (c) $u = -D\frac{\partial J}{\partial x}$

   (d) $u = D\frac{\partial J}{\partial x}$

4. The term D in the diffusion equation represents what?

   (a) Density

   (b) Viscosity

   (c) Diffusion coefficient

   (d) Heat capacity

5. What numerical method is commonly used to solve the diffusion equation when analytical solutions are not available?

   (a) Finite difference method

   (b) Finite element method

   (c) Spectral method

   (d) Boundary integral method

6. Which of the following conditions is crucial for the stability of numerical methods in solving the diffusion equation?

   (a) CFL condition

   (b) Nyquist theorem

   (c) Root mean square error

   (d) Convergence criterion

7. In what context is the diffusion equation often applied in biological processes?

   (a) Population dynamics

   (b) Genetic drift

   (c) Oxygen transport in tissues

(d) Neural signal transmission

Answers:

1. **C: Conservation of mass** The diffusion equation is derived from conservation of mass principles, specifically concerning the transfer and distribution of substances in space over time.

2. **A:** $\frac{\partial u(x,t)}{\partial t} = D\frac{\partial^2 u(x,t)}{\partial x^2}$ This is the standard form of the one-dimensional diffusion equation, representing the time evolution of a concentration field governed by spatial second derivatives.

3. **B:** $J = -D\frac{\partial u}{\partial x}$ Fick's law states that the flux J of a diffusing substance is proportional to the negative gradient of the concentration u, with D being the diffusion coefficient.

4. **C: Diffusion coefficient** The term D in the diffusion equation quantifies how rapidly the diffusive process occurs, characterizing the efficiency of substance transfer.

5. **B: Finite element method** The finite element method is widely used for solving the diffusion equation numerically, especially when dealing with complex geometries and varying boundary conditions.

6. **A: CFL condition** The Courant-Friedrichs-Lewy (CFL) condition is essential in numerical methods to ensure stability, particularly in explicit time-stepping schemes for parabolic equations like the diffusion equation.

7. **C: Oxygen transport in tissues** The diffusion equation applies to biological systems where it describes processes such as oxygen and nutrient transport in living tissues, an essential aspect for understanding physiological functions.

Chapter 6

Navier-Stokes Equations for Incompressible Flow

The Navier-Stokes equations for incompressible flow play a central role in the study of fluid dynamics. This chapter delves into the fundamental principles underlying these equations and explores various mathematical and computational techniques used to analyze and solve them. We begin by introducing the underlying principles of fluid mechanics and establishing the necessary mathematical framework. Then, we derive the Navier-Stokes equations using conservation laws and discuss their physical and mathematical significance. Next, we explore different approaches to formulating and solving these equations, including the weak formulation, vorticity-stream function approach, and computational fluid dynamics (CFD) methods. Finally, we present real-world applications of the Navier-Stokes equations in diverse fields, showcasing their broad impact and relevance.

Fundamental Principles

Fluid mechanics is the branch of physics that studies the properties and behavior of fluids, both in motion and at rest. It provides a framework for understanding the motion, forces, and interactions of fluids, which are essential in various scientific and engineering disciplines. The study of fluid mechanics relies on various fundamental principles, including conservation of mass, conservation of momentum, and the constitutive relationships that describe the behavior of fluids.

Derivation and Physical Context

The Navier-Stokes equations describe the motion of an incompressible fluid based on conservation principles. They are derived by applying the principles of conservation of mass and conservation of momentum to a fluid element. The resulting equations mathematically express the time evolution of the fluid velocity, pressure, and other relevant variables. The Navier-Stokes equations consider both the external forces acting on the fluid and the internal viscous forces resulting from fluid flow.

Weak Formulation

The Navier-Stokes equations, being nonlinear partial differential equations, can be challenging to solve directly. To overcome this difficulty, a common approach is to reformulate the problem in a weak form using variational principles. This weak form leads to a set of integral equations, which can be readily solved using numerical methods. The weak formulation provides a flexible framework for the numerical approximation of fluid flow problems.

Vorticity-Stream Function Approach

The vorticity-stream function formulation is an alternative mathematical approach to solving the Navier-Stokes equations. In this approach, the vorticity, a measure of the local rotation of the fluid, is used as the primary variable, along with the stream function, which is related to the flow velocity components. This formulation simplifies the equations and allows for more efficient computational techniques, particularly for two-dimensional, steady, and irrotational flows.

Computational Fluid Dynamics Methods

Computational fluid dynamics (CFD) methods are powerful tools used to solve the Navier-Stokes equations numerically. These methods discretize the fluid domain into a computational grid and solve the governing equations using iterative techniques. Various numerical schemes, such as finite difference, finite volume, and finite element methods, can be employed to obtain accurate approximations of the fluid flow behavior. CFD methods have revolutionized the study of fluid dynamics by enabling detailed analyses of complex flow phenomena.

Real-World Applications

The Navier-Stokes equations find extensive applications in a wide range of real-world scenarios. They are used to model and analyze fluid flow in many fields, including aerospace engineering, environmental science, geophysics, and biomedical engineering. Examples of applications include studying aircraft aerodynamics, analyzing ocean currents, simulating blood flow in the human body, and designing efficient cooling systems. The Navier-Stokes equations provide a fundamental framework for understanding and predicting fluid flow phenomena in diverse practical contexts.

In summary, the Navier-Stokes equations are fundamental to the study of fluid dynamics, providing a mathematical description of the motion and behavior of incompressible fluids. By utilizing various mathematical and computational methods, researchers and engineers can analyze and solve these equations to gain insights into a wide range of fluid flow problems. The Navier-Stokes equations have proven invaluable in understanding complex fluid dynamic phenomena and have facilitated advancements in numerous scientific and engineering fields.

Python Code Snippet

Below is a Python code snippet that implements key algorithms and equations related to the Navier-Stokes equations for incompressible fluid flow. This code includes implementations of the weak formulation using finite difference methods and the computation of the velocity field using the vorticity-stream function approach.

```python
import numpy as np
import matplotlib.pyplot as plt

def initialize_grid(nx, ny, Lx, Ly):
    '''
    Initialize the grid for simulation.
    :param nx: Number of grid points in the x direction.
    :param ny: Number of grid points in the y direction.
    :param Lx: Physical size of the domain in the x direction.
    :param Ly: Physical size of the domain in the y direction.
    :return: x and y grid coordinates.
    '''
    x = np.linspace(0, Lx, nx)
    y = np.linspace(0, Ly, ny)
    return np.meshgrid(x, y)

def navier_stokes_step(u, v, dt, dx, dy, nu):
    '''
    Perform one time step of the Navier-Stokes update.
    :param u: Velocity field in the x direction.
    :param v: Velocity field in the y direction.
    :param dt: Time step size.
    :param dx: Space step size in the x direction.
    :param dy: Space step size in the y direction.
    :param nu: Kinematic viscosity.
    :return: Updated velocity fields u and v.
    '''
    un = u.copy()
    vn = v.copy()

    # Poisson's equation for pressure
    b = (un[1:-1, 2:] - un[1:-1, :-2])/(2*dx) + (vn[2:, 1:-1] -
        vn[:-2, 1:-1])/(2*dy)
    p = np.zeros_like(un)

    # Solve for pressure (simple finite difference method)
    p[1:-1, 1:-1] = p[1:-1, 1:-1] + (b[1:-1, :-2] - b[1:-1,
        2:])/(2*dx) + (b[:-2, 1:-1] - b[2:, 1:-1])/(2*dy)

    # Update velocities
    u[1:-1, 1:-1] = (un[1:-1, 1:-1] - un[1:-1, 1:-1] * dt/dx *
        (un[1:-1, 1:-1] - un[1:-1, :-2])
                    - vn[1:-1, 1:-1] * dt/dy * (un[1:-1, 1:-1] -
                        un[:-2, 1:-1])
                    - (p[1:-1, 2:] - p[1:-1, :-2]) * dt/(2*dx)
                    + nu * (un[1:-1, 2:] - 2*un[1:-1, 1:-1] +
                        un[1:-1, :-2]) * dt/dx**2
                    + nu * (un[2:, 1:-1] - 2*un[1:-1, 1:-1] +
                        un[:-2, 1:-1]) * dt/dy**2)

    v[1:-1, 1:-1] = (vn[1:-1, 1:-1] - un[1:-1, 1:-1] * dt/dx *
        (vn[1:-1, 1:-1] - vn[1:-1, :-2])
```

```
                    - vn[1:-1, 1:-1] * dt/dy * (vn[1:-1, 1:-1] -
                      vn[:-2, 1:-1])
                    - (p[2:, 1:-1] - p[:-2, 1:-1]) * dt/(2*dy)
                    + nu * (vn[1:-1, 2:] - 2*vn[1:-1, 1:-1] +
                      vn[1:-1, :-2]) * dt/dx**2
                    + nu * (vn[2:, 1:-1] - 2*vn[1:-1, 1:-1] +
                      vn[:-2, 1:-1]) * dt/dy**2)

        return u, v

# Set parameters
nx, ny = 50, 50  # Grid size
Lx, Ly = 2.0, 2.0  # Domain size
nu = 0.1  # Kinematic viscosity
dt = 0.01  # Time step
nt = 100  # Number of time steps

# Initialize velocity fields
u = np.zeros((nx, ny))
v = np.zeros((nx, ny))

# Create grid
x, y = initialize_grid(nx, ny, Lx, Ly)

# Time-stepping loop for solving the Navier-Stokes equations
for n in range(nt):
    u, v = navier_stokes_step(u, v, dt, Lx/nx, Ly/ny, nu)

# Visualization
plt.quiver(x, y, u, v)
plt.title('Velocity Field After Simulation')
plt.xlabel('X-axis')
plt.ylabel('Y-axis')
plt.show()
```

This code defines two primary functions:

- `initialize_grid` initializes a uniform grid for the computational domain.
- `navier_stokes_step` computes one time step of the Navier-Stokes equations using finite difference methods for updating velocity and pressure fields.

The simulation parameters, including grid size and kinematic viscosity, can be adjusted to explore different fluid flow scenarios. After running the simulation, the final velocity field is visualized using a quiver plot to illustrate the flow dynamics in the computational domain.

Multiple Choice Questions

1. Which principle is primarily used to derive the Navier-Stokes equations?

    (a) Conservation of mass

    (b) Conservation of momentum

    (c) Conservation of energy

    (d) Both conservation of mass and momentum

2. What does the term "incompressible" indicate in the context of fluid flow?

    (a) The fluid density remains constant

    (b) The fluid velocity is constant

    (c) The pressure of the fluid remains constant

    (d) The temperature of the fluid remains constant

3. In the weak formulation of the Navier-Stokes equations, which of the following is primarily sought?

    (a) A classical solution

    (b) An integral equation form

    (c) A unique solution with boundary conditions

    (d) An asymptotic solution

4. The vorticity-stream function approach is particularly advantageous for what kind of flows?

    (a) Unsteady three-dimensional flows

    (b) Steady and irrotational flows

    (c) Two-dimensional flows

    (d) Compressible flows

5. Which of the following methods is a common numerical approach used in Computational Fluid Dynamics (CFD)?

    (a) Finite difference method

    (b) Finite element method

    (c) Spectral method

(d) All of the above

6. Real-world applications of the Navier-Stokes equations include which of the following?

    (a) Modeling weather patterns
    (b) Analyzing ocean currents
    (c) Simulating blood flow in the body
    (d) All of the above

7. True or False: The Navier-Stokes equations can only be solved analytically.

    (a) True
    (b) False

Answers:

1. **D: Both conservation of mass and momentum** The Navier-Stokes equations are derived by applying both the principles of conservation of mass (continuity equation) and conservation of momentum to fluid elements. These principles combine to describe the dynamics of incompressible flow.

2. **A: The fluid density remains constant** Incompressible flow means that the density of the fluid does not change with variations in pressure or temperature, simplifying the analysis of fluid movement.

3. **B: An integral equation form** The weak formulation of the Navier-Stokes equations converts the problem into an integral form that is more manageable for numerical methods, particularly for non-linear PDEs.

4. **C: Two-dimensional flows** The vorticity-stream function formulation is beneficial primarily for two-dimensional flows, allowing for easier numerical implementations and conservation of properties.

5. **D: All of the above** All mentioned methods—finite difference, finite element, and spectral methods—are established numerical techniques employed in CFD to solve the Navier-Stokes equations.

6. **D: All of the above** The Navier-Stokes equations are widely applicable in areas such as weather forecasting, oceanography, and biomedical engineering, which all demand fluid dynamics analysis.

7. **B: False** While many Navier-Stokes problems are complex and not solvable analytically, numerous numerical methods exist that allow for approximate solutions. This is crucial because many real-world fluid flow problems can only be solved numerically.

Chapter 7

Maxwell's Equations in Electromagnetics

In this chapter, we explore Maxwell's equations in the field of electromagnetics. These equations form the foundation of classical electrodynamics and describe the fundamental principles of electric and magnetic fields. We begin by discussing the system of equations and their basic concepts. Then, we delve into the time-harmonic case, which is of particular interest in studying electromagnetic waves. We also introduce boundary-integral methods as an approach to solving Maxwell's equations, followed by an examination of finite element methods as an alternative numerical technique. Lastly, we discuss applications of Maxwell's equations in the field of telecommunications.

System of Equations and Basic Concepts

Maxwell's equations are a set of partial differential equations that govern the behavior of electric and magnetic fields in the presence of sources. They are based on four fundamental laws: Gauss's law for electric fields, Gauss's law for magnetic fields, Faraday's law of electromagnetic induction, and Ampère's law with Maxwell's addition. Together, these equations provide a complete description of the interaction between electric and magnetic fields, as well as their generation and propagation.

Mathematically, Maxwell's equations can be written as follows:

$$\nabla \cdot \mathbf{E} = \frac{\rho}{\varepsilon_0} \quad \text{(Gauss's law for electric fields)}$$

$$\nabla \cdot \mathbf{B} = 0 \quad \text{(Gauss's law for magnetic fields)}$$

$$\nabla \times \mathbf{E} = -\frac{\partial \mathbf{B}}{\partial t} \quad \text{(Faraday's law of electromagnetic induction)}$$

$$\nabla \times \mathbf{B} = \mu_0 \mathbf{J} + \mu_0 \varepsilon_0 \frac{\partial \mathbf{E}}{\partial t}$$
$$\text{(Ampère's law with Maxwell's addition)}$$

Here, **E** and **B** represent the electric and magnetic fields, respectively. ρ is the charge density, **J** is the current density, ε_0 is the vacuum permittivity, and μ_0 is the vacuum permeability. These equations reflect the conservation of electric charge, the absence of magnetic monopoles, the generation of electric fields by time-varying magnetic fields, and the generation of magnetic fields by electric currents and time-varying electric fields.

Time-Harmonic Case

In many electromagnetics applications, such as studying electromagnetic waves, it is common to consider the time-harmonic case. This involves assuming that the electric and magnetic fields can be represented as complex phasors with a specific angular frequency, typically denoted as ω. In this case, Maxwell's equations can be written in the frequency domain as:

$$\nabla \cdot \mathbf{E} = \frac{\rho}{\varepsilon_0}$$
$$\nabla \cdot \mathbf{B} = 0$$
$$\nabla \times \mathbf{E} = -j\omega \mathbf{B}$$
$$\nabla \times \mathbf{B} = j\omega \mu_0 \mathbf{J} + j\omega \mu_0 \varepsilon_0 \mathbf{E}$$

Here, j represents the imaginary unit, and the time dependence of the fields is given by $e^{j\omega t}$. The time-harmonic case simplifies the analysis and enables efficient numerical methods for solving electromagnetic problems.

Boundary-Integral Methods

Boundary-integral methods provide an alternative approach to solving Maxwell's equations, particularly for problems with complex geometries and boundary conditions. These methods convert the volume integral equations of Maxwell's equations into equivalent surface integral equations over the boundaries of the computational domain. By discretizing the boundary surfaces and solving the resulting integral equations, one can obtain accurate solutions for various electromagnetic problems.

Finite Element Methods

Finite element methods (FEM) have also proven effective in solving Maxwell's equations numerically. FEM involves discretizing the computational domain into a set of smaller subdomains called finite elements. The electric and magnetic fields are then approximated within each element using interpolation functions. By solving the resulting algebraic system, which arises from enforcing Maxwell's equations within each element and across element boundaries, one can obtain numerical solutions for the electric and magnetic fields.

Applications in Telecommunications

Maxwell's equations find numerous applications in the field of telecommunications. These include the design and analysis of antennas, waveguides, transmission lines, and microwave components. By solving Maxwell's equations, researchers and engineers can predict the behavior of electromagnetic waves, optimize the performance of communication systems, and ensure electromagnetic compatibility. Maxwell's equations form the basis for electromagnetic simulators and computer-aided design tools widely used in the telecommunications industry.

In conclusion, Maxwell's equations provide a comprehensive mathematical framework for understanding and analyzing the behavior of electric and magnetic fields. By solving these equations, researchers and engineers gain insights into electromagnetic phenomena, enabling advancements in various fields. Applications in telecommunications harness the power of Maxwell's equations in the design and optimization of communication systems. Through the use of numerical methods and specialized techniques, Maxwell's equations continue to be a cornerstone of electromagnetic theory and its practical applications.

Python Code Snippet

Below is a Python code snippet that implements key equations and techniques discussed in the chapter on Maxwell's equations in electromagnetics. This code includes the formulation for calculating the electric and magnetic fields in the time-harmonic case, as well as the methods for numerical solutions using Finite Element Methods (FEM).

```python
import numpy as np
import matplotlib.pyplot as plt

def calculate_electric_field(charge_density, permittivity, x, y):
    '''
    Calculate the electric field produced by a point charge
    ↪   distribution.
    :param charge_density: Charge density (C/m^3).
    :param permittivity: Permittivity of the medium (F/m).
    :param x: x-coordinates of the field points.
    :param y: y-coordinates of the field points.
    :return: Electric field (Ex, Ey) components.
    '''
    # Compute total electric field using Gauss's law
    Ex = charge_density / permittivity * np.ones(x.shape)
    Ey = charge_density / permittivity * np.ones(y.shape)
    return Ex, Ey

def calculate_magnetic_field(current_density, permeability, x, y):
    '''
    Calculate the magnetic field produced by a current density
    ↪   distribution.
    :param current_density: Current density (A/m^2).
    :param permeability: Permeability of the medium (H/m).
    :param x: x-coordinates of the field points.
    :param y: y-coordinates of the field points.
    :return: Magnetic field (Bx, By) components.
    '''
    # Compute total magnetic field using Ampère's law
    Bx = permeability * current_density * np.ones(x.shape)
    By = permeability * current_density * np.ones(y.shape)
    return Bx, By

def plot_fields(x, y, Ex, Ey, Bx, By):
    '''
```

```python
    Plot the electric and magnetic fields.
    :param x: x-coordinates of the field points.
    :param y: y-coordinates of the field points.
    :param Ex: Electric field x-component.
    :param Ey: Electric field y-component.
    :param Bx: Magnetic field x-component.
    :param By: Magnetic field y-component.
    '''
    plt.figure(figsize=(12, 6))

    # Plot Electric field
    plt.subplot(1, 2, 1)
    plt.quiver(x, y, Ex, Ey, color='r', scale=5)
    plt.title("Electric Field")
    plt.xlabel("x (m)")
    plt.ylabel("y (m)")
    plt.axis('equal')

    # Plot Magnetic field
    plt.subplot(1, 2, 2)
    plt.quiver(x, y, Bx, By, color='b', scale=5)
    plt.title("Magnetic Field")
    plt.xlabel("x (m)")
    plt.ylabel("y (m)")
    plt.axis('equal')

    plt.tight_layout()
    plt.show()

# Inputs for the calculations
charge_density = 1e-6  # Charge density in C/m^3
current_density = 1e-3  # Current density in A/m^2
permittivity = 8.854187e-12  # Permittivity of free space in F/m
permeability = 1.256637e-6  # Permeability of free space in H/m

# Create grid points
x = np.linspace(-1, 1, 20)
y = np.linspace(-1, 1, 20)
X, Y = np.meshgrid(x, y)

# Calculate fields
Ex, Ey = calculate_electric_field(charge_density, permittivity, X,
↪   Y)
Bx, By = calculate_magnetic_field(current_density, permeability, X,
↪   Y)

# Plotting the fields
plot_fields(X, Y, Ex, Ey, Bx, By)
```

This code defines three functions:

- `calculate_electric_field` computes the electric field from a

given charge density and permittivity based on Gauss's law.
- `calculate_magnetic_field` calculates the magnetic field using a current density and permeability based on Ampère's law.
- `plot_fields` visualizes the computed electric and magnetic fields using quiver plots for clear representation of vector fields.

The example provided calculates the electric and magnetic fields in a uniform distribution from specified charge and current densities, and then plots the results to visually represent the fields.

Multiple Choice Questions

1. Which of the following equations represents Gauss's law for electric fields?

    (a) $\nabla \cdot \mathbf{B} = 0$
    (b) $\nabla \times \mathbf{E} = -\frac{\partial \mathbf{B}}{\partial t}$
    (c) $\nabla \cdot \mathbf{E} = \frac{\rho}{\varepsilon_0}$
    (d) $\nabla \times \mathbf{B} = \mu_0 \mathbf{J} + \mu_0 \varepsilon_0 \frac{\partial \mathbf{E}}{\partial t}$

2. In the time-harmonic case of Maxwell's equations, how is the time dependence of the fields typically represented?

    (a) $e^{\omega t}$
    (b) $e^{-j\omega t}$
    (c) $j\omega t$
    (d) $e^{j\omega t}$

3. What is the primary purpose of boundary-integral methods in solving Maxwell's equations?

    (a) To simplify the volume of the problem
    (b) To solve the equations using approximations
    (c) To convert volume integrals into surface integrals
    (d) To increase the dimensionality of the problem

4. Which numerical method is based on discretizing the computational domain into smaller subdomains called finite elements?

    (a) Boundary-Integral Method
    (b) Finite Element Method

(c) Finite Difference Method

(d) Spectral Method

5. Which of the following is NOT an application of Maxwell's equations in telecommunications?

    (a) Antenna design

    (b) Waveguides

    (c) Economic model predictions

    (d) Microwave components

6. What does the term "electric displacement field" (often referred to as **D**) relate to in the context of Maxwell's equations?

    (a) The voltage in a circuit

    (b) The electric field and charge density

    (c) The intensity of magnetic fields

    (d) The potential energy in a field

7. What type of condition can be applied to the electromagnetic fields at the interface of two different media?

    (a) Boundary Conditions

    (b) Initial Conditions

    (c) Stability Conditions

    (d) Convergence Conditions

Answers:

1. **C:** $\nabla \cdot \mathbf{E} = \frac{\rho}{\varepsilon_0}$ This equation represents Gauss's law for electric fields, which relates the divergence of the electric field to the charge density, indicating how electric charges create electric fields.

2. **D:** $e^{j\omega t}$ In the time-harmonic case, the electric and magnetic fields are represented as oscillating sinusoidal functions, which is typically expressed in the form $e^{j\omega t}$, where ω is the angular frequency.

3. **C: To convert volume integrals into surface integrals** The primary purpose of boundary-integral methods is to reformulate Maxwell's equations, allowing for more efficient handling of complex geometries by transforming volume integrals into surface integrals over the domain's boundary.

4. **B: Finite Element Method** The Finite Element Method (FEM) involves breaking down a computational domain into smaller elements and is a widely used numerical technique for solving partial differential equations, including Maxwell's equations.

5. **C: Economic model predictions** This option does not relate to electromagnetic theory or applications of Maxwell's equations. In contrast, antenna design, waveguides, and microwave components all directly involve electromagnetic concepts.

6. **B: The electric field and charge density** The electric displacement field, denoted by **D**, relates to the electric field **E** and the free charge density in the context of dielectric materials, incorporating effects of polarization.

7. **A: Boundary Conditions** Boundary conditions are crucial in electromagnetic problems as they dictate how fields behave at the interface between different media, ensuring that both the electric and magnetic fields satisfy the physical laws dictated by Maxwell's equations.

Chapter 8

The Cauchy-Riemann Equations

The Cauchy-Riemann Equations lie at the heart of complex analysis, providing a crucial connection between complex-differentiability and the existence of analytic functions. In this chapter, we delve into the foundation of complex analysis, starting with an introduction to the basic concepts of complex numbers and their representation in the complex plane. We then present the mathematical formulation of the Cauchy-Riemann equations and explore their implications. Additionally, we explore the solutions of these equations in polar coordinates and examine numerical implementations. Lastly, we discuss various applications of the Cauchy-Riemann equations in fluid dynamics.

Introduction to Complex Analysis

Complex analysis is the branch of mathematics that extends the theory of calculus to complex numbers. A complex number is of the form $z = x + iy$, where x and y are real numbers and i is the imaginary unit. The complex plane provides a geometric representation of complex numbers, with the real axis corresponding to the x-axis and the imaginary axis corresponding to the y-axis.

1 Representation in the Complex Plane

In the complex plane, complex numbers can be expressed in polar form as $z = r \cdot e^{i\theta}$, where $r = |z|$ is the modulus or absolute value of z, and θ is the argument or angle of z. This representation allows for a concise and convenient way to represent complex numbers and perform operations such as multiplication and exponentiation.

2 Analytic Functions

An analytic function is a complex-valued function that is differentiable in a region of the complex plane. The concept of complex differentiability is closely tied to the Cauchy-Riemann equations and plays a central role in complex analysis. An analytic function has a power series representation in terms of its derivatives, known as a Taylor series, which enables the evaluation of its values and derivatives at any point within its domain.

Mathematical Formulation

The Cauchy-Riemann equations establish a necessary condition for the differentiability of a function in terms of its partial derivatives. For a complex function $f(z) = u(x,y) + iv(x,y)$ to be complex-differentiable at a point $z = x + iy$, the following conditions must hold:

$$\frac{\partial u}{\partial x} = \frac{\partial v}{\partial y} \tag{1}$$

$$\frac{\partial u}{\partial y} = -\frac{\partial v}{\partial x} \tag{2}$$

Here, $u(x,y)$ and $v(x,y)$ are the real and imaginary parts of $f(z)$, respectively. Equations (1) and (2) are known as the Cauchy-Riemann equations and provide a powerful tool for analyzing the differentiability and the existence of an analytic function.

Polar Coordinates Solutions

Solutions to the Cauchy-Riemann equations can also be expressed in polar coordinates, providing insights into the behavior of functions in terms of their magnitude and phase. By substituting

$z = re^{i\theta}$ and expressing u and v as functions of r and θ, the Cauchy-Riemann equations can be rewritten as follows:

$$\frac{\partial u}{\partial r} = \frac{1}{r}\frac{\partial v}{\partial \theta} \tag{3}$$

$$\frac{\partial u}{\partial \theta} = -r\frac{\partial v}{\partial r} \tag{4}$$

These equations provide a convenient framework for studying functions with radial and angular symmetries.

Numerical Implementations

Numerically solving the Cauchy-Riemann equations can be challenging due to the presence of partial derivatives. Various numerical methods, such as finite difference and finite element, can be employed to approximate the derivatives and solve the equations within a computational domain. These methods involve discretizing the domain and expressing the derivatives as algebraic equations, which can be solved using numerical techniques.

Applications in Fluid Dynamics

The Cauchy-Riemann equations find notable applications in fluid dynamics, particularly in the study of potential flows. Potential flows are described by a complex potential function, from which fluid velocity, pressure, and streamlines can be derived. The satisfaction of the Cauchy-Riemann equations for the complex potential ensures that the flow is irrotational and incompressible, providing a simplified yet insightful analysis framework.

The Cauchy-Riemann equations also play a role in the analysis of conformal mappings, which preserve shape and angles. By satisfying the Cauchy-Riemann equations, a complex function can generate conformal maps, enabling the transformation of geometric shapes with minimal distortion. Conformal mappings have various applications in fluid dynamics, such as the analysis of airfoils and the study of fluid flow around objects.

In conclusion, the Cauchy-Riemann equations form the backbone of complex analysis and provide essential insights into the differentiability and existence of analytic functions. They establish a connection between complex differentiability and the satisfaction of partial derivative conditions. Solutions to these equations can be obtained in both Cartesian and polar coordinates, allowing for a thorough investigation of complex functions. Numerical implementations are employed to solve these equations in computational domains, while applications in fluid dynamics, particularly in potential flows and conformal mappings, showcase their significance in diverse areas of research.

Python Code Snippet

Below is a Python code snippet that demonstrates how to solve the Cauchy-Riemann equations and compute potential functions in the context of fluid dynamics.

```python
import numpy as np
import matplotlib.pyplot as plt

def cauchy_riemann_solve(u, v, x_range, y_range, num_points=100):
    '''
    Solves the Cauchy-Riemann equations given the real part u and
    ↪ imaginary part v on specified ranges.
    :param u: Callable function representing the real part of the
    ↪ complex function.
    :param v: Callable function representing the imaginary part of
    ↪ the complex function.
    :param x_range: Tuple specifying the range of x values (xmin,
    ↪ xmax).
    :param y_range: Tuple specifying the range of y values (ymin,
    ↪ ymax).
    :param num_points: Number of points in each direction for the
    ↪ grid.
    :return: Grid of values for x, y, u, and v.
    '''
    x = np.linspace(x_range[0], x_range[1], num_points)
    y = np.linspace(y_range[0], y_range[1], num_points)
    X, Y = np.meshgrid(x, y)

    U = u(X, Y)
    V = v(X, Y)

    return X, Y, U, V


def example_u(x, y):
    '''
    Example function of u(x, y) = x^2 - y^2
    '''
    return x**2 - y**2
```

```
def example_v(x, y):
    '''
    Example function of v(x, y) = 2xy
    '''
    return 2 * x * y


# Define ranges for x and y
x_range = (-2, 2)
y_range = (-2, 2)

# Solve Cauchy-Riemann equations
X, Y, U, V = cauchy_riemann_solve(example_u, example_v, x_range,
 ↪ y_range)

# Plotting the results
plt.figure(figsize=(10, 8))
plt.quiver(X, Y, U, V, color='blue')
plt.title('Vector Field from Cauchy-Riemann Functions')
plt.xlabel('x')
plt.ylabel('y')
plt.xlim(x_range)
plt.ylim(y_range)
plt.axhline(0, color='black',linewidth=0.5, ls='--')
plt.axvline(0, color='black',linewidth=0.5, ls='--')
plt.grid()
plt.show()
```

This code defines several functions to demonstrate the solutions of the Cauchy-Riemann equations in a computational domain:

- cauchy_riemann_solve computes the grid of values for the real part $u(x, y)$ and imaginary part $v(x, y)$ given the ranges for x and y.
- example_u represents a specific form of the real part, defined as $u(x, y) = x^2 - y^2$.
- example_v represents a corresponding imaginary part, defined as $v(x, y) = 2xy$.

The provided example generates a vector field from the functions satisfying the Cauchy-Riemann equations and visualizes this field using a quiver plot. This illustrates how the functions function in the context of complex analysis and fluid dynamics, showing the relationship between potential functions and their derivatives as specified by the equations.

Multiple Choice Questions

1. What is the form of a complex number in the complex plane?
   (a) $z = a + b$
   (b) $z = r\cos(\theta) + ir\sin(\theta)$
   (c) $z = x + iy$, where $x, y \in \mathbb{R}$
   (d) $z = x^2 + iy^2$

2. Which of the following statements regarding analytic functions is TRUE?
   (a) Analytic functions are only defined on the real line.
   (b) Analytic functions can be expressed as a polynomial of any degree.
   (c) Analytic functions are complex-differentiable in a neighborhood of every point in their domain.
   (d) Analytic functions cannot have derivatives at any point.

3. The Cauchy-Riemann equations provide conditions for which property of complex functions?
   (a) Continuity
   (b) Differentiability
   (c) Integrability
   (d) Monotonicity

4. In polar coordinates, the Cauchy-Riemann equations can be expressed as:
   (a) $\frac{\partial u}{\partial r} = \frac{1}{r}\frac{\partial v}{\partial \theta}$
   (b) $\frac{\partial u}{\partial r} = r\frac{\partial v}{\partial \theta}$
   (c) $\frac{\partial v}{\partial r} = \frac{\partial u}{\partial \theta}$
   (d) $\frac{\partial u}{\partial r} = -\frac{\partial v}{\partial \theta}$

5. Which of the following numerical methods can be applied to solve the Cauchy-Riemann equations?
   (a) Monte Carlo Simulation
   (b) Discrete Fourier Transform

    (c) Finite Difference Method

    (d) Kalman Filter

6. The applications of the Cauchy-Riemann equations in fluid dynamics primarily relate to:

    (a) Compressible flows

    (b) Steady-state heat transfer

    (c) Potential flows and the concept of ideal fluids

    (d) Turbulent flows

7. True or False: The satisfaction of the Cauchy-Riemann equations implies that a function is analytic everywhere in the complex plane.

    (a) True

    (b) False

Answers:
1. **C:** $z = x + iy$, **where** $x, y \in \mathbb{R}$
This option correctly reflects the general representation of complex numbers in Cartesian coordinates, where x is the real part and y is the imaginary part of the complex number z.

2. **C: Analytic functions are complex-differentiable in a neighborhood of every point in their domain.**
Analytic functions are defined as functions that are differentiable not only at a point but also in a neighborhood around that point, which implies a higher degree of smoothness.

3. **B: Differentiability**
The Cauchy-Riemann equations establish the necessary conditions for a function to be differentiable in the complex sense, indicating that the function must satisfy certain relations between its real and imaginary parts.

4. **A:** $\frac{\partial u}{\partial r} = \frac{1}{r}\frac{\partial v}{\partial \theta}$
This option correctly states one of the forms of the Cauchy-Riemann equations when expressed in polar coordinates, relating radial and angular derivatives of the real and imaginary parts.

5. **C: Finite Difference Method**
The finite difference method is a common numerical approach used to approximate derivatives, making it applicable for solving the Cauchy-Riemann equations through discretization.

6. C: Potential flows and the concept of ideal fluids
The Cauchy-Riemann equations are significantly relevant for describing potential flows in fluid dynamics, where fluid motion can be analyzed using a complex potential function that satisfies these equations.

7. B: False
While the satisfaction of the Cauchy-Riemann equations is a necessary condition for a function to be analytic, it is not sufficient alone for all functions unless specified that this holds in a neighborhood. A function can satisfy the Cauchy-Riemann equations at isolated points and still not be analytic.

Chapter 9
Schrödinger Equation

In this chapter, we delve into the Schrödinger equation, a fundamental equation in quantum mechanics that describes the behavior of quantum systems. Named after Erwin Schrödinger, the equation provides insight into the wave-particle duality of quantum objects and allows us to study the behavior and properties of microscopic particles such as electrons and atoms. In this chapter, we explore the background of quantum mechanics, the mathematical formulation of the Schrödinger equation, and its various solutions and applications.

Quantum Mechanics Background

Quantum mechanics is a branch of physics that addresses the behavior and interactions of particles at the microscopic level. It was developed in the early 20th century to explain phenomena that classical mechanics and classical electromagnetism could not account for. Quantum mechanics is built on the principles of superposition and wave-particle duality, which postulate that particles can exist in multiple states simultaneously and can exhibit both wave-like and particle-like properties.

Mathematical Formulation

The Schrödinger equation is a partial differential equation that describes the evolution of the wave function of a quantum system over time. It is given by:

$$i\hbar\frac{\partial \Psi}{\partial t} = -\frac{\hbar^2}{2m}\nabla^2\Psi + V\Psi$$

where Ψ is the wave function of the system, t is time, $\hbar$ is the reduced Planck constant, m is the mass of the particle, ∇^2 is the Laplacian operator, and V is the potential energy function. The Schrödinger equation is a complex-valued equation, with the solutions providing the probability amplitudes of different states that the quantum system can occupy.

Analytical Solutions for Simple Potentials

Analytical solutions to the Schrödinger equation are only possible for a limited number of potential energy functions. For simple potentials, such as the harmonic oscillator potential and the infinite square well potential, analytical solutions can be found. These solutions often involve solving a second-order ordinary differential equation or applying separation of variables techniques.

For example, the time-independent Schrödinger equation for the one-dimensional harmonic oscillator potential is given by:

$$-\frac{\hbar^2}{2m}\frac{d^2\psi}{dx^2} + \frac{1}{2}m\omega^2 x^2 \psi = E\psi$$

where $\psi(x)$ is the wave function, x is the spatial coordinate, ω is the angular frequency of the oscillator, and E is the energy eigenvalue. The solutions to this equation are given by Hermite polynomials multiplied by a Gaussian factor.

Spectral Methods

In cases where analytical solutions are not feasible, numerical methods such as spectral methods can be employed to approximate solutions to the Schrödinger equation. Spectral methods involve expanding the wave function in terms of a set of basis functions and determining the expansion coefficients using techniques like the Fourier series or the Chebyshev polynomials. These methods provide accurate approximations of the wave function and the corresponding eigenvalues.

Solving Time-Dependent Schrödinger Equation

The time-dependent Schrödinger equation allows us to study the time evolution of a quantum system and the time dependence of its wave function. This equation is given by:

$$i\hbar \frac{\partial \Psi}{\partial t} = \hat{H}\Psi$$

where $\hat{H}$ is the Hamiltonian operator, representing the total energy of the system. Solving the time-dependent Schrödinger equation requires finding the eigenfunctions and eigenvalues of the Hamiltonian operator, which can be challenging for complex systems. Numerical methods, such as finite difference or split-operator methods, are often used to obtain approximate solutions.

Quantum Computing Applications

The Schrödinger equation is of particular importance in the field of quantum computing. Quantum computers leverage the principles of quantum mechanics to perform certain computations much more efficiently than classical computers. The wave function and its evolution are central to the operation of quantum algorithms, and the Schrödinger equation forms the foundation for understanding and designing quantum algorithms. The ability to manipulate and control quantum systems based on the solutions to the Schrödinger equation enables the development of powerful quantum technologies.

In conclusion, the Schrödinger equation provides a fundamental framework for understanding the behavior of quantum systems and the wave-like properties of particles. The equation's mathematical formulation allows us to study the time evolution and energy states of quantum systems, though analytical solutions are limited to simple potentials. Numerical methods and spectral methods enable the approximation of solutions for more complex systems. The Schrödinger equation plays a crucial role in the field of quantum computing, where it underpins the design and operation of quantum algorithms and quantum technologies.

Python Code Snippet

Below is a Python code snippet that implements important equations and algorithms related to the Schrödinger equation, including analytical solutions for a harmonic oscillator potential, the time-dependent Schrödinger equation, and spectral methods for approximating the wave function.

```python
import numpy as np
import matplotlib.pyplot as plt
from scipy.integrate import solve_ivp
from scipy.special import hermite

def harmonic_oscillator_wavefunction(n, x):
    '''
    Calculate the wavefunction for the n-th state of a
        one-dimensional harmonic oscillator.
    :param n: Quantum number (0, 1, 2, ...).
    :param x: Spatial coordinate.
    :return: The wavefunction value.
    '''
    normalization = (m * omega / (np.pi * hbar))**0.25
    factor = normalization * (1 / np.sqrt(2**n *
        np.math.factorial(n)))
    polynomial = hermite(n)(np.sqrt(m * omega / hbar) * x)
    return factor * np.exp(-m * omega * x**2 / (2 * hbar)) *
        polynomial

def time_evolution(n, t, x):
    '''
    Calculate the time evolution of the wavefunction for the n-th
        state.
    :param n: Quantum number (0, 1, 2, ...).
    :param t: Time variable.
    :param x: Spatial coordinate.
    :return: The wavefunction value at time t.
    '''
    energy = (n + 0.5) * hbar * omega
    return harmonic_oscillator_wavefunction(n, x) * np.exp(-1j *
        energy * t / hbar)

def solve_time_dependent_schrodinger(n, x, t):
    '''
    Solve the time-dependent Schrödinger equation for a quantum
        harmonic oscillator
    and return the wavefunction at all time points.
    :param n: Quantum number (0, 1, 2, ...).
    :param x: Array of spatial coordinates.
    :param t: Array of time points.
    :return: The wavefunction for each time point.
    '''
```

```
        wavefunctions = []
        for time in t:
            wavefunction = time_evolution(n, time, x)
            wavefunctions.append(wavefunction)
        return np.array(wavefunctions)

# Constants
hbar = 1.0545718e-34    # Reduced Planck constant (J·s)
m = 9.10938356e-31      # Mass of the electron (kg)
omega = 1.0             # Angular frequency (rad/s)

# Parameters for the quantum harmonic oscillator
n = 0                              # Ground state
x_values = np.linspace(-5, 5, 400)  # Spatial coordinates
t_values = np.linspace(0, 10, 100)  # Time points

# Solve the time-dependent Schrödinger equation
wavefunctions = solve_time_dependent_schrodinger(n, x_values,
↪   t_values)

# Plotting the wavefunctions at different time points
plt.figure(figsize=(10, 6))
for index, time in enumerate([0, 2, 4, 6, 8]):
    plt.plot(x_values, np.abs(wavefunctions[index])**2, label=f'Time
↪   = {time}s')
plt.title('Time Evolution of the Ground State Wavefunction')
plt.xlabel('Position (x)')
plt.ylabel('Probability Density |(x,t)|²')
plt.legend()
plt.grid()
plt.show()
```

This code defines several functions:

- harmonic_oscillator_wavefunction calculates the wave function for the n-th state of a one-dimensional harmonic oscillator based on the quantum number, spatial coordinate, mass, and frequency.
- time_evolution determines the time-evolved wave function for the n-th state given the time and position.
- solve_time_dependent_schrodinger solves the time-dependent Schrödinger equation for a quantum harmonic oscillator and returns the wave function at all specified time points.

The provided example calculates the time evolution of the wave function for the ground state of a harmonic oscillator and plots the probability density at different time moments. This visual representation helps in understanding the dynamics of quantum states as they evolve over time.

Multiple Choice Questions

1. What is the physical significance of the wave function Ψ in quantum mechanics?

    (a) It represents the position of a particle at a given time.

    (b) It provides the probability amplitude for the outcomes of measurements.

    (c) It describes the exact trajectory of a particle.

    (d) It is an observable quantity that can be directly measured.

2. In the time-independent Schrödinger equation, which of the following terms represents the potential energy?

    (a) $i\hbar \frac{\partial \Psi}{\partial t}$

    (b) $-\frac{\hbar^2}{2m}\nabla^2 \Psi$

    (c) $V\Psi$

    (d) Ψ

3. What type of boundary conditions are typically necessary to solve the Schrödinger equation?

    (a) Homogeneous boundary conditions only

    (b) Non-homogeneous boundary conditions only

    (c) Mixed boundary conditions

    (d) There are no boundary conditions required

4. Which equation allows for the analysis of how a quantum system evolves over time?

    (a) Time-independent Schrödinger equation

    (b) Heisenberg uncertainty principle

    (c) Time-dependent Schrödinger equation

    (d) Pauli exclusion principle

5. What roles do Hermite polynomials play in the context of the one-dimensional harmonic oscillator?

    (a) They determine the time evolution of the wave function.

(b) They are the solutions to the time-dependent Schrödinger equation.

(c) They form part of the solutions to the time-independent Schrödinger equation.

(d) They represent the potential energy in the harmonic oscillator.

6. In quantum computing, what aspect of the Schrödinger equation is crucial for performing computations?

    (a) The linearity of the equation allows for superposition states.

    (b) The complexity of calculations needed to derive eigenvalues.

    (c) The use of wave functions as classical objects.

    (d) The direct measurement of wave functions.

7. Which of the following methods is commonly used to numerically solve the Schrödinger equation for complex potentials?

    (a) Singular value decomposition

    (b) Fourier transforms

    (c) Finite difference methods

    (d) Laplace transforms

Answers:

1. **B: It provides the probability amplitude for the outcomes of measurements.** The wave function Ψ contains all the information about a quantum system, and its absolute square $|\Psi|^2$ gives the probability density for finding a particle in a particular state or position.

2. **C:** $V\Psi$ In the time-independent Schrödinger equation, the term $V\Psi$ represents the potential energy of the system, indicating how potential energy influences the behavior of the wave function.

3. **C: Mixed boundary conditions** Solving the Schrödinger equation usually requires boundary conditions that can be mixed, depending on the potential or the physical constraints of the problem.

4. **C: Time-dependent Schrödinger equation** The time-dependent Schrödinger equation describes how the wave function of a quantum system evolves over time, allowing analysis of dynamic quantum systems.

5. **C: They form part of the solutions to the time-independent Schrödinger equation.** Hermite polynomials are part of the analytical solutions to the time-independent Schrödinger equation for the harmonic oscillator, contributing to the bound state wave functions.

6. **A: The linearity of the equation allows for superposition states.** The linear nature of the Schrödinger equation enables quantum systems to exist in superposition states, which is fundamental to the operation and efficiency of quantum algorithms in quantum computing.

7. **C: Finite difference methods** Finite difference methods are commonly used to numerically approximate solutions to the Schrödinger equation, especially when dealing with complex potentials where analytical solutions are not feasible.

Chapter 10

The Monge-Ampère Equation

Geometric Background

The Monge-Ampère equation is a fully nonlinear elliptic partial differential equation that has important applications in geometry and optimization. It is named after Gaspard Monge and André-Marie Ampère, who made significant contributions to its study. The Monge-Ampère equation is a fundamental tool in various fields, including geometric optics, optimal transportation, and differential geometry.

In geometric terms, the Monge-Ampère equation arises in the study of optimal mass transport and optimal mappings between two domains. It quantifies the distortion or the change of volume between two regions. This equation plays a crucial role in understanding the properties and shape-preserving transformations of various geometric objects.

Weak Solutions

The Monge-Ampère equation can be formulated as a second-order fully nonlinear equation of the form:

$$(\det D^2 u)(x) = f(x)$$

where u is the unknown function, $D^2 u$ is the Hessian matrix of second partial derivatives of u, and f is a given function. The equation involves the determinant of the Hessian matrix, representing the change of volume or distortion of the mapping defined by u.

Finding a solution to the Monge-Ampère equation is challenging due to its nonlinearity. Weak solutions, also known as viscosity solutions, provide a viable approach for solving the equation in a generalized sense. Weak solutions relax the requirement of differentiability and allow for solutions that satisfy the equation in a distributional sense.

Numerical Solutions via Finite Differences

Numerical methods, particularly finite difference methods, are widely used to approximate solutions to the Monge-Ampère equation. In the finite difference approach, the domain is discretized into a grid of points, and difference approximations are used to discretize the derivatives and the equation itself. The resulting system of algebraic equations can then be solved iteratively.

One popular numerical scheme for solving the Monge-Ampère equation is the *Douglas-Rachford method*. This method combines a convex splitting technique with finite differences to approximate the equation efficiently.

Convex Optimization Algorithms

Convex optimization algorithms, such as the *interior point method* and the *primal-dual method of multipliers*, have proven to be effective in solving the Monge-Ampère equation. These algorithms exploit the convexity of the equation and can handle nonlinear constraints. They provide efficient and robust numerical solutions for various practical applications.

Applications in Geometric Optics

The Monge-Ampère equation finds important applications in geometric optics, a branch of optics concerned with the study of light propagation in media with varying refractive indices. The equation describes the refracting properties of curved surfaces and allows for the design and optimization of optical systems with desired optical properties. The Monge-Ampère equation appears in problems such as lens design, image formation, and ray tracing.

In conclusion, the Monge-Ampère equation is a powerful and versatile tool in geometry and optimization. Its geometric interpretation and applications in various fields make it a central topic of study. Weak solutions and numerical methods, including finite difference techniques and convex optimization algorithms, provide effective means for solving and approximating solutions to this nonlinear equation. In particular, the Monge-Ampère equation finds practical applications in geometric optics, where it plays a key role in designing optical systems and analyzing light propagation in refractive media. "'latex

Python Code Snippet

Below is a Python code snippet that implements important algorithms and numerical methods for solving the Monge-Ampère equation, including the Douglas-Rachford method and a basic implementation of convex optimization techniques.

```python
import numpy as np

def douglas_rachford_method(f, u0, alpha, max_iterations, tol):
    '''
    The Douglas-Rachford method for solving the Monge-Ampère
    ↪ equation.
    :param f: The function on the right side of the Monge-Ampère
    ↪ equation.
    :param u0: Initial guess for the solution.
    :param alpha: Step size in the iterative process.
    :param max_iterations: Maximum number of iterations.
    :param tol: Tolerance for stopping criterion.
    :return: Approximated solution.
    '''
    u = u0.copy()
    for iteration in range(max_iterations):
        u_prev = u.copy()

        # First half step (forward step)
        gradient = np.gradient(u)
        u_half = u - alpha * np.linalg.det(np.gradient(u))

        # Second half step (backward step)
        u = 0.5 * (u + u_half - alpha *
            np.linalg.det(np.gradient(u_half)))

        # Check convergence
```

```python
        if np.linalg.norm(u - u_prev) < tol:
            break

    return u


def convex_optimization_method(f, u0, max_iterations, tol):
    '''
    Convex optimization technique for minimizing the Monge-Ampère
    ↪ equation.
    :param f: Function to be optimized.
    :param u0: Initial guess for the solution.
    :param max_iterations: Maximum number of iterations.
    :param tol: Tolerance for stopping criterion.
    :return: Optimized result.
    '''
    u = u0.copy()
    for iteration in range(max_iterations):
        u_prev = u.copy()

        # Compute the gradient and Hessian
        gradient = np.gradient(u)
        hessian = np.linalg.hess(u)

        # Update step (gradient descent)
        u -= 0.1 * gradient  # Fixed small step to minimize the
        ↪ function

        # Project onto the convex set if necessary

        # Check convergence
        if np.linalg.norm(u - u_prev) < tol:
            break

    return u


# Parameters and inputs
mesh_size = 100
x = np.linspace(0, 1, mesh_size)
y = np.linspace(0, 1, mesh_size)
X, Y = np.meshgrid(x, y)

f = np.exp(-((X - 0.5)**2 + (Y - 0.5)**2))  # Example right-hand
↪ side function
u0 = np.zeros_like(f)  # Initial guess
alpha = 0.01  # Step size for Douglas-Rachford method
max_iterations = 1000  # Limit for iterations
tol = 1e-5  # Tolerance level

# Solve using Douglas-Rachford method
u_approx = douglas_rachford_method(f, u0, alpha, max_iterations,
↪ tol)
```

```
# Solve using convex optimization method
u_optimal = convex_optimization_method(f, u0, max_iterations, tol)

# Output results
print("Approximate solution using Douglas-Rachford Method:",
 ↪ u_approx)
print("Optimal solution using Convex Optimization Method:",
 ↪ u_optimal)
```

This code defines two functions:

- `douglas_rachford_method` implements the Douglas-Rachford method to approximate the solution of the Monge-Ampère equation.
- `convex_optimization_method` provides a basic convex optimization technique to minimize the function associated with the equation.

The example illustrates how to set parameters, initialize values, and apply both methods to approximate solutions to the Monge-Ampère equation, printing the resulting outputs for comparison.
"'

Multiple Choice Questions

1. Which of the following best describes the Monge-Ampère equation?

    (a) A linear equation involving first-order derivatives

    (b) A linear equation involving second-order derivatives

    (c) A fully nonlinear elliptic partial differential equation

    (d) A hyperbolic equation governing wave propagation

2. The term "weak solution" in the context of the Monge-Ampère equation refers to:

    (a) A solution that is defined only at the boundary

    (b) A solution that satisfies the equation in a distributional sense

    (c) A solution that is approximated using numerical methods

    (d) A solution with discontinuities

3. Which numerical method is commonly used to approximate solutions to the Monge-Ampère equation?

    (a) Finite Element Method

    (b) Finite Difference Method

    (c) Method of Characteristics

    (d) Boundary Integral Method

4. What role does the determinant of the Hessian matrix play in the Monge-Ampère equation?

    (a) It represents the rate of change of the function

    (b) It measures the curvature of the function

    (c) It signifies the distortion or change of volume in the mapping

    (d) It defines the boundary conditions for the solution

5. In which field does the Monge-Ampère equation have notable applications?

    (a) Financial mathematics

    (b) Quantum mechanics

    (c) Geometric optics

    (d) Game theory

6. Which optimization algorithm is mentioned as being effective for solving the Monge-Ampère equation?

    (a) Simplex Method

    (b) Gradient Descent Algorithm

    (c) Interior Point Method

    (d) Genetic Algorithm

7. True or False: The Monge-Ampère equation can be used to optimize the design of optical systems.

    (a) True

    (b) False

Answers:

1. **C: A fully nonlinear elliptic partial differential equation** The Monge-Ampère equation is indeed categorized as a fully nonlinear elliptic PDE, distinguishing it from linear equations.

2. **B: A solution that satisfies the equation in a distributional sense** In the context of nonlinear equations, weak solutions (or viscosity solutions) generalize the concept of a solution by allowing for less regularity.

3. **B: Finite Difference Method** The finite difference method is a widely used technique for approximating solutions to PDEs, including the Monge-Ampère equation.

4. **C: It signifies the distortion or change of volume in the mapping** The determinant of the Hessian in the Monge-Ampère equation represents how the volume changes as a result of the mapping defined by the function u.

5. **C: Geometric optics** The Monge-Ampère equation has critical applications in geometric optics, particularly concerning the design and optimization of optical systems.

6. **C: Interior Point Method** The interior point method is a powerful convex optimization technique that has been effectively applied to solving the Monge-Ampère equation.

7. **A: True** The Monge-Ampère equation is indeed instrumental in optimizing the design of optical systems by describing the refracting properties of surfaces.

Chapter 11

The Black-Scholes Equation

Introduction

In this chapter, we delve into the mathematical formulation, derivation, and solution techniques associated with the Black-Scholes equation. This equation is a pivotal model in the field of financial mathematics, specifically in the pricing of derivative securities and option contracts. Developed by economists Fischer Black and Myron Scholes in 1973, the Black-Scholes equation revolutionized options pricing by introducing a mathematical framework to model the dynamics of stock prices and derive fair prices for options.

Financial Derivatives and Option Pricing

Financial derivatives are contracts whose values derive from the underlying assets, such as stocks, commodities, or currencies. Options are a type of financial derivative that provide the right, but not the obligation, to buy (call option) or sell (put option) an underlying asset at a specified price (strike price) within a certain time period (expiration date).

Option pricing is a crucial aspect of financial markets, enabling investors to assess the fair value of options and make informed investment decisions. The Black-Scholes equation provides a mathematical framework for option pricing by modeling the dynamics of stock prices and determining the fair price of options based on key parameters like volatility, interest rates, and time to expiration.

Derivation and Assumptions

The Black-Scholes equation is derived under several assumptions, which include:

1. The underlying stock price follows a geometric Brownian motion.

2. The market is efficient and does not exhibit arbitrage opportunities.

3. Continuous trading is allowed, and there are no transaction costs.

4. The risk-free interest rate is constant over the life of the option.

5. The returns on the underlying asset are normally distributed.

Under these assumptions, the Black-Scholes equation provides a mathematical representation of the relationship between the price of the option, the underlying asset price, and other parameters.

Analytical and Semi-Analytical Solutions

The Black-Scholes equation is a partial differential equation (PDE) that relates the fair price of an option to various factors, including the underlying asset price, volatility, time to expiration, strike price, and the risk-free interest rate. While the Black-Scholes equation does not have a closed-form solution for most option contracts, analytical and semi-analytical solutions are available for specific cases.

1 Closed-Form Solutions

The Black-Scholes equation has closed-form solutions for European call options and put options. The Black-Scholes formula, derived from the equation, provides explicit formulas to calculate these option prices. The formula for the price of a European call option is given by:

$$C = S_0 N(d_1) - Xe^{-rT} N(d_2)$$

where: C is the call option price, S_0 is the current stock price, $N(\cdot)$ is the cumulative distribution function of the standard normal distribution, X is the strike price, r is the risk-free interest rate, T is the time to expiration, and d_1 and d_2 are defined as:

$$d_1 = \frac{\ln\left(\frac{S_0}{X}\right) + (r + \frac{1}{2}\sigma^2)T}{\sigma\sqrt{T}}$$

$$d_2 = d_1 - \sigma\sqrt{T}$$

2 Numerical Solutions

For more complex option contracts or when closed-form solutions are not available, numerical methods are employed to solve the Black-Scholes equation. Finite difference methods and Monte Carlo simulations are commonly used techniques to obtain numerical approximations for option prices.

Finite difference methods discretize the Black-Scholes equation and approximate the derivatives using finite differences. By iterating through time and space, the option prices at each computational point are obtained. These methods can handle a wide range of option contracts, but they require careful consideration of stability and convergence.

Monte Carlo simulations simulate the underlying asset price paths using random sampling techniques and estimate option prices through statistical analysis. These methods provide a flexible framework for pricing complex options, but they can be computationally intensive.

Practical Applications in Finance

The Black-Scholes equation and its pricing framework have extensive applications in finance. Some notable applications include:

1. Option Pricing: The Black-Scholes equation forms the foundation for pricing options, aiding in determining fair prices for call and put options across various asset classes.

2. **Risk Management:** Option prices obtained from the Black-Scholes model help market participants assess and manage risk exposures associated with financial derivatives.

3. **Portfolio Management:** The Black-Scholes model supports portfolio optimization by identifying mispriced options and constructing optimal investment strategies.

4. **Volatility Estimation:** The Black-Scholes equation provides a method to estimate implied volatility, which is a critical input for option pricing and risk management.

In conclusion, the Black-Scholes equation is a fundamental tool in option pricing and has revolutionized the field of financial mathematics. Through its derivation and assumptions, it allows for the determination of fair prices for options based on various factors. While the equation has closed-form solutions for specific option contracts, numerical methods play a crucial role in approximating option prices for more complex contracts. The practical applications of the Black-Scholes equation extend beyond option pricing and contribute to risk management, portfolio optimization, and volatility estimation in financial markets.

Python Code Snippet

Below is a Python code snippet that implements the important equations and algorithms related to the Black-Scholes equation for option pricing. It includes functions to calculate the price of European call options, handle numerical simulations via finite differences, and estimate implied volatility.

```python
import numpy as np
import scipy.stats as si

def black_scholes_call(S, K, T, r, sigma):
    '''
    Calculate the price of a European call option using the
       Black-Scholes formula.
    :param S: Current stock price.
    :param K: Strike price of the option.
    :param T: Time to expiration in years.
    :param r: Risk-free interest rate (annual).
    :param sigma: Volatility of the underlying stock (annualized).
    :return: Call option price.
    '''
```

```python
    # Calculate d1 and d2
    d1 = (np.log(S / K) + (r + 0.5 * sigma ** 2) * T) / (sigma *
    ↪ np.sqrt(T))
    d2 = d1 - sigma * np.sqrt(T)

    # Calculate call option price using the cumulative distribution
    ↪ function
    call_price = S * si.norm.cdf(d1) - K * np.exp(-r * T) *
    ↪ si.norm.cdf(d2)
    return call_price


def finite_difference_black_scholes(S, K, T, r, sigma, Smax, M, N):
    '''
    Calculate the price of a European call option using finite
    ↪ difference methods.
    :param S: Current stock price.
    :param K: Strike price of the option.
    :param T: Time to expiration in years.
    :param r: Risk-free interest rate (annual).
    :param sigma: Volatility of the underlying stock (annualized).
    :param Smax: Maximum stock price considered in the grid.
    :param M: Number of asset price steps.
    :param N: Number of time steps.
    :return: Price of the European call option using finite
    ↪ difference methods.
    '''
    # Create the price and time grid
    S_values = np.linspace(0, Smax, M)
    time_grid = np.linspace(0, T, N)

    V = np.maximum(S_values - K, 0)  # Payoff at maturity for
    ↪ European call option

    # Coefficients for finite difference method
    dt = T / N
    dS = Smax / M
    alpha = 0.5 * dt * ((sigma ** 2) * (np.arange(M) ** 2) - r *
    ↪ np.arange(M))
    beta = 1 - dt * (sigma ** 2 * np.arange(M) ** 2 + r)
    gamma = 0.5 * dt * ((sigma ** 2) * (np.arange(M) ** 2) + r)

    # Time-stepping backward
    for j in range(N-1, 0, -1):
        for i in range(1, M-1):
            V[i] = alpha[i] * V[i-1] + beta[i] * V[i] + gamma[i] *
            ↪ V[i+1]
        V[0] = 0  # Boundary Condition: V(0) = 0
        V[-1] = (Smax - K * np.exp(-r * (T - time_grid[j]))) *
        ↪ np.exp(-r * (T - time_grid[j]))  # Boundary Condition:
        ↪ V(Smax)
```

```python
        return V[int(M/2)]  # Return call price corresponding to current
                            # stock price


def implied_volatility(S, K, T, r, market_price, tol=1e-5,
                       max_iterations=100):
    '''
    Calculate the implied volatility using the Newton-Raphson
        method.
    :param S: Current stock price.
    :param K: Strike price of the option.
    :param T: Time to expiration in years.
    :param r: Risk-free interest rate (annual).
    :param market_price: Market price of the option.
    :param tol: Tolerance for convergence.
    :param max_iterations: Maximum number of iterations.
    :return: Implied volatility.
    '''
    # Initial guess for volatility
    sigma = 0.2

    for i in range(max_iterations):
        price = black_scholes_call(S, K, T, r, sigma)
        vega = S * si.norm.pdf((np.log(S / K) + (r + 0.5 * sigma **
                2) * T) / (sigma * np.sqrt(T))) * np.sqrt(T)

        # Update guess for sigma
        sigma -= (price - market_price) / vega

        if abs(price - market_price) < tol:
            return sigma

    return sigma  # Return final sigma

# Inputs for the calculations
S = 100     # Current stock price
K = 100     # Strike price
T = 1       # Time to expiration in years
r = 0.05    # Risk-free interest rate
sigma = 0.2 # Initial volatility
Smax = 200  # Maximum stock price for finite difference method
M = 100     # Number of asset price steps
N = 100     # Number of time steps
market_price = 10   # Market price for implied volatility calculation

# Calculations
call_price = black_scholes_call(S, K, T, r, sigma)
fd_call_price = finite_difference_black_scholes(S, K, T, r, sigma,
        Smax, M, N)
iv = implied_volatility(S, K, T, r, market_price)

# Output results
print("Black-Scholes Call Option Price:", call_price)
```

```
print("Finite Difference Call Option Price:", fd_call_price)
print("Implied Volatility:", iv)
```

This code includes three primary functions:

- `black_scholes_call` computes the price of a European call option using the Black-Scholes formula.
- `finite_difference_black_scholes` employs finite difference methods to calculate the price of a European call option.
- `implied_volatility` estimates the implied volatility based on the market price of the option using the Newton-Raphson method.

The provided example demonstrates how to calculate option prices and implied volatility, printing the results for analysis.

Multiple Choice Questions

1. Who were the primary developers of the Black-Scholes equation?

    (a) Robert Merton and Eugene Fama

    (b) Fischer Black and Myron Scholes

    (c) John Hull and Richard Black

    (d) Paul Samuelson and William Sharpe

2. What is the core assumption regarding the stock price in the Black-Scholes model?

    (a) The stock price is constant over time.

    (b) The stock price follows a geometric Brownian motion.

    (c) The stock price fluctuates based on market sentiment.

    (d) The stock price is determined by supply and demand.

3. Which of the following best describes a European option?

    (a) It can be exercised at any time before expiration.

    (b) It can only be exercised on the expiration date.

    (c) It has no expiration date.

    (d) It can only be exercised during specific trading hours.

4. The term "implied volatility" in the context of the Black-Scholes model refers to:

(a) The actual volatility of the stock.

(b) The volatility used in the model to arrive at an observed option price.

(c) Historical volatility calculated from past stock prices.

(d) The volatility expected from future market conditions.

5. Which of the following methodologies is NOT typically used to solve the Black-Scholes equation numerically?

   (a) Finite difference methods

   (b) Monte Carlo simulations

   (c) Backward induction

   (d) Analytical solutions

6. In the Black-Scholes formula, what does the term $N(d_1)$ represent?

   (a) The risk-free interest rate

   (b) The cumulative distribution function for a standard normal variable

   (c) The current stock price

   (d) The option's strike price

7. The Black-Scholes equation has crucial implications for which of the following financial practices?

   (a) Arbitrage

   (b) Risk management and hedging

   (c) Regulatory compliance

   (d) Tax optimization strategies

Answers:

1. **B: Fischer Black and Myron Scholes** Fischer Black and Myron Scholes developed the Black-Scholes equation in 1973, fundamentally changing the way options are priced in financial markets.

2. **B: The stock price follows a geometric Brownian motion.** The Black-Scholes model assumes that the underlying stock price evolves according to geometric Brownian motion, reflecting both the random walk nature of stock prices and the continuous compounding of returns.

3. **B: It can only be exercised on the expiration date.** European options can only be exercised at expiration as opposed to American options, which can be exercised at any time before or on the expiration date.

4. **B: The volatility used in the model to arrive at an observed option price.** Implied volatility is the market's forecast of the underlying asset's volatility, derived from the market price of an option, reflecting how volatile investors expect the underlying asset to be in the future.

5. **D: Analytical solutions** While analytical solutions provide exact prices for standard options, they are not considered numerical methods. Finite difference methods and Monte Carlo simulations are commonly used for numerical solutions when closed forms are unavailable.

6. **B: The cumulative distribution function for a standard normal variable** The term $N(d_1)$ in the Black-Scholes formula refers to the cumulative distribution function of the standard normal distribution, which helps in determining probabilities related to stock price movements.

7. **B: Risk management and hedging** The Black-Scholes equation is essential for risk management and hedging, as it provides a framework for pricing options, allowing market participants to design strategies that manage exposure to price movements in underlying assets.

Chapter 12

Elliptic Regularity Theory

Sobolev Spaces

In this section, we introduce the concept of Sobolev spaces, which are a fundamental tool in the study of elliptic partial differential equations (PDEs) and their regularity. Sobolev spaces provide a framework for analyzing the smoothness and regularity properties of functions by considering their weak derivatives. These spaces are equipped with appropriate norms, allowing for the formulation of well-posedness results for elliptic PDEs and the study of their solutions.

1 Definition

Let Ω be a bounded domain in $\mathbb{R}^n$ with a smooth boundary. The Sobolev space $W^{k,p}(\Omega)$, where $k \in \mathbb{N}$ and $p \geq 1$, is defined as the set of functions $u \in L^p(\Omega)$ for which all weak derivatives up to order k belong to $L^p(\Omega)$. More formally, we have

$$W^{k,p}(\Omega) = \{u \in L^p(\Omega) : D^\alpha u \in L^p(\Omega) \text{ for all } |\alpha| \leq k\},$$

where $D^\alpha u$ denotes the weak derivative of u of order $\alpha = (\alpha_1, \alpha_2, \ldots, \alpha_n)$.

2 Norm and Completeness

The Sobolev space $W^{k,p}(\Omega)$ is endowed with the norm

$$\|u\|_{W^{k,p}(\Omega)} = \left(\sum_{|\alpha|\leq k} \int_\Omega |D^\alpha u|^p\, dx\right)^{1/p},$$

which measures the L^p-integrability of the weak derivatives of order up to k. It can be shown that this norm defines a complete normed vector space. In particular, the Sobolev space $W^{k,p}(\Omega)$ is a Banach space.

3 Sobolev Embedding Theorems

One of the key features of Sobolev spaces is their embedding properties, which relate the smoothness of functions to their regularity in terms of integrability. These embedding theorems establish the inclusion relationships between Sobolev spaces of different orders and integrability classes. Two important embedding theorems are:

Sobolev Embedding Theorem (Compact Embedding): Let Ω be a bounded domain with a smooth boundary and $p \geq n$. Then, there exists a constant $C > 0$ such that for any $u \in W^{k,p}(\Omega)$ with $k > \frac{n}{p}$, we have

$$\|u\|_{L^q(\Omega)} \leq C\|u\|_{W^{k,p}(\Omega)},$$

where $q = \frac{np}{n-kp}$.

Sobolev Embedding Theorem (Sobolev-Slobodeckii Embedding): Let Ω be a bounded domain with a smooth boundary and $p > n$. Then, there exists a constant $C > 0$ such that for any $u \in W^{k,p}(\Omega)$ with $k > \frac{n}{p}$, we have

$$\|u\|_{C^{m,\alpha}(\overline{\Omega})} \leq C\|u\|_{W^{k,p}(\Omega)},$$

where m is any nonnegative integer, $\alpha \in (0,1)$, and $C^{m,\alpha}(\overline{\Omega})$ denotes the space of functions whose derivatives up to order m are in $C^\alpha(\overline{\Omega})$.

These embedding theorems are powerful tools that establish the regularity properties of solutions to elliptic PDEs and provide important estimates for the existence and uniqueness of solutions.

Weak Solutions & Lax-Milgram Theorem

In this section, we discuss the notion of weak solutions for elliptic partial differential equations (PDEs) and the Lax-Milgram theorem, which provides a general framework for establishing the existence and uniqueness of weak solutions.

1 Definition of Weak Solutions

Consider the general elliptic PDE of the form

$$Lu = -\sum_{i,j=1}^{n} \frac{\partial}{\partial x_i}\left(a^{ij}\frac{\partial u}{\partial x_j}\right) + \sum_{i=1}^{n} b^i \frac{\partial u}{\partial x_i} + cu = f,$$

where L is a second-order linear differential operator with smooth coefficients a^{ij}, b^i, and c, and f is a given function. A weak solution to this PDE is a function $u \in W^{1,2}(\Omega)$, where Ω is the domain of interest, that satisfies the weak formulation

$$\int_\Omega a^{ij} \frac{\partial u}{\partial x_j}\frac{\partial \phi}{\partial x_i} + \int_\Omega b^i \frac{\partial u}{\partial x_i}\phi + \int_\Omega cu\phi = \int_\Omega f\phi,$$

for all test functions $\phi \in W_0^{1,2}(\Omega)$, where $W_0^{1,2}(\Omega)$ is the subspace of $W^{1,2}(\Omega)$ consisting of functions that vanish on the boundary of Ω.

2 Lax-Milgram Theorem

The Lax-Milgram theorem is a fundamental result in the theory of elliptic PDEs that establishes the existence and uniqueness of weak solutions under appropriate assumptions.

Lax-Milgram Theorem: Let V be a Hilbert space, $a(\cdot,\cdot)$ a bounded, coercive bilinear form on V, and F a bounded linear functional on V. Then, there exists a unique solution $u \in V$ to the problem

$$a(u,v) = F(v), \quad \forall v \in V.$$

In the context of elliptic PDEs, the Lax-Milgram theorem provides a rigorous framework for demonstrating the existence and uniqueness of weak solutions to elliptic boundary value problems using variational methods. By properly choosing the function spaces and establishing appropriate coercivity and boundedness properties, the Lax-Milgram theorem plays a central role in the study of elliptic partial differential equations.

3 Applications

The theory of weak solutions and the Lax-Milgram theorem find wide-ranging applications in various areas of mathematics and engineering. Some notable applications include:

- **Boundary Value Problems:** The theory of weak solutions allows for the treatment of non-smooth domains or boundary conditions in the analysis of boundary value problems for elliptic PDEs.

- **Finite Element Methods:** Weak solutions form the basis for the formulation and analysis of finite element methods, which are widely used numerical methods for solving PDEs in complex domains.

- **Optimal Control Theory:** The existence and uniqueness of weak solutions are crucial for the study of optimal control problems governed by elliptic PDEs, enabling the development of efficient algorithms for control and optimization.

- **Stability Analysis:** Weak solutions and the Lax-Milgram theorem provide tools for investigating the stability and convergence of numerical schemes and approximation methods for elliptic PDEs.

These applications highlight the fundamental role of weak solutions and the Lax-Milgram theorem in the analysis, modeling, and numerical approximation of elliptic PDEs, making them essential tools for researchers and engineers in various disciplines.

Schauder Estimates

In this section, we discuss Schauder estimates, which provide crucial bounds on the regularity of solutions to elliptic partial differential equations (PDEs). Schauder estimates establish the Hölder continuity, and thus the smoothness, of solutions by relating the regularity of the coefficients in the PDE and the regularity of the right-hand side.

1 Statement of Schauder Estimates

Consider the general elliptic PDE

$$Lu = -\sum_{i,j=1}^{n} \frac{\partial}{\partial x_i}\left(a^{ij}\frac{\partial u}{\partial x_j}\right) + \sum_{i=1}^{n} b^i \frac{\partial u}{\partial x_i} + cu = f,$$

where L is a second-order linear differential operator with smooth coefficients a^{ij}, b^i, and c, and f is a given function. The Schauder estimates provide bounds on the Hölder norm of the solution u in terms of the Hölder norm of the coefficients and the right-hand side.

Schauder Estimate: Let Ω be a domain with Hölder continuous boundary, $0 < \alpha \leq 1$, and $f \in C^\beta(\Omega)$ for some $\beta > 0$. Assume that the coefficients a^{ij}, b^i, and c are Hölder continuous, i.e., $a^{ij}, b^i, c \in C^\gamma(\overline{\Omega})$ for some $\gamma > 0$. Then, the solution u of the elliptic PDE satisfies

$$\|u\|_{C^{2,\alpha}(\overline{\Omega})} \leq C\left(\|f\|_{C^\beta(\Omega)} + \sum_{|\alpha|\leq 2, |\beta|\leq 2} \|D^\alpha a^{ij}\|_{C^\gamma(\overline{\Omega})} \|D^\beta u\|_{C^\gamma(\overline{\Omega})}\right),$$

where C depends on Ω, α, β, and γ, but is independent of u and f.

The Schauder estimates are a powerful tool in the regularity theory of elliptic PDEs as they quantify the regularity of solutions and provide control on the smoothness of solutions based on the regularity of the coefficients and the right-hand side. These estimates have significant implications for the well-posedness, numerical approximation, and stability analysis of elliptic PDEs.

2 Practical Significance

Schauder estimates have practical significance in various fields of mathematics, physics, and engineering. Some notable applications include:

- **Boundary Value Problems:** The Schauder estimates establish the regularity of solutions to boundary value problems for elliptic PDEs, allowing for the analysis of physical phenomena in diverse fields, including fluid mechanics, heat transfer, and electromagnetics.

- **Numerical Analysis:** The Schauder estimates are used to analyze the convergence and accuracy of numerical methods, such as finite element methods, finite difference methods, and spectral methods, for the solution of elliptic PDEs.

- **Optimal Control:** By providing bounds on the regularity of solutions to elliptic PDEs, Schauder estimates are instrumental in the theoretical and numerical analysis of optimal control problems, optimization, and inverse problems.

- **Homogenization and Multiscale Problems:** The regularity estimates afforded by Schauder estimates are crucial in the study of homogenization and multiscale problems, where the solutions exhibit oscillations and variations on multiple length scales.

These applications highlight the profound influence of Schauder estimates on the understanding and analysis of elliptic PDEs, allowing for the formulation of accurate mathematical models and the development of efficient numerical algorithms in numerous scientific and engineering disciplines.

3 Extensions and Refinements

The original Schauder estimates have been extended and refined in various directions to address specific classes of elliptic PDEs and to obtain sharper estimates for higher regularity. For instance, when the coefficients of the equation are more regular, such as Hölder continuous or analytic, the Schauder estimates yield stronger regularity results for the solutions. Moreover, adaptations of Schauder estimates have been developed for more general elliptic systems, nonlinear equations, and non-smooth domains. These extensions and refinements contribute to a deeper understanding of the regularity theory of elliptic PDEs and enable more precise analysis of specific classes of problems of practical interest.

Python Code Snippet

Below is a Python code snippet that implements important equations and algorithms mentioned in this chapter, including the calculation of Sobolev norms, weak solution verification via the Lax-Milgram theorem, and Schauder estimates for elliptic PDEs.

```python
import numpy as np
from scipy.integrate import quad

def sobolev_norm(u, k, p, domain):
    '''
    Calculate the Sobolev norm of a function u in Sobolev space
    ↪ W^{k,p} over a given domain.
    :param u: Function (as a callable) for which Sobolev norm is to
    ↪ be computed.
    :param k: Order of weak derivatives to consider.
    :param p: Integrability class.
    :param domain: Tuple representing the limits of integration (a,
    ↪ b).
    :return: Sobolev norm.
    '''
    norm = 0
    # Compute the L^p norm for derivatives up to order k
    for order in range(k + 1):
        # Define the derivative
        derivative = np.polyder(u, order)
        # Calculate integral of |derivative|^p
        norm += quad(lambda x: abs(derivative(x))**p, *domain)[0]

    return (norm)**(1/p)


def weak_solution_test(a, b, f, phi, domain):
    '''
    Evaluate the weak formulation for a given function.
    :param a: Coefficient function a^{ij}.
    :param b: Coefficient function b^i.
    :param f: Source term function f.
    :param phi: Test function phi.
    :param domain: Tuple representing the limits of integration (a,
    ↪ b).
    :return: Left-hand side of the weak formulation integral.
    '''
    lhs = 0
    for i in range(len(a)):
        lhs += quad(lambda x: a[i] * np.gradient(phi(x)),
            ↪ *domain)[0]

    lhs += quad(lambda x: b * phi(x), *domain)[0]
    lhs += quad(lambda x: f * phi(x), *domain)[0]

    return lhs


def schauder_estimate(u, coeffs, p, domain):
    '''
    Estimate the regularity of the solution using Schauder
    ↪ estimates.
```

```python
    :param u: Function (as a callable) representing the solution.
    :param coeffs: Coefficients including a^{ij}, b^i, c.
    :param p: Hölder continuity exponent.
    :param domain: Tuple representing the limits of integration (a,
    ↪    b).
    :return: Estimate of the regularity of the solution in
    ↪    C^{2,\alpha} norm.
    '''
    # Here we assume coeffs is a dictionary containing 'a', 'b', and
    ↪    'c'
    estimate = 0
    for a in coeffs['a']:
        estimate += np.max(np.abs(a))   # Estimating based on max
        ↪    coefficients

    estimate *= (1 + np.linalg.norm(u(domain[0]:domain[1]))) ** p

    return estimate


# Example input functions
def example_function(x):
    return x**2

# Coefficients as an example (simple case)
coeffs = {
    'a': [1.0, 0.5],    # example coefficient for a^{ij}
    'b': 2.0,           # example coefficient for b^i
    'c': 1.0            # constant term
}

# Domain of integration for Sobolev space
domain = (0, 1)

# Calculating Sobolev norm
sobolev_norm_result = sobolev_norm(example_function, 2, 2, domain)

# Verifying weak solution
phi_test = lambda x: x   # Simple test function
weak_form_result = weak_solution_test(coeffs['a'], coeffs['b'],
↪    example_function, phi_test, domain)

# Schauder estimate for regularity assessment
schauder_result = schauder_estimate(example_function, coeffs, 0.5,
↪    domain)

# Output results
print("Sobolev Norm Result:", sobolev_norm_result)
print("Weak Solution Test Result:", weak_form_result)
print("Schauder Estimate Result:", schauder_result)
```

This code defines the following functions:

- `sobolev_norm` computes the Sobolev norm of a given function up to order k and for Lebesgue space L^p.
- `weak_solution_test` evaluates the left-hand side of the weak formulation of a PDE given coefficient functions, a source term, and a test function.
- `schauder_estimate` provides an estimate for the regularity of a solution based on given coefficients and Hölder continuity.

The provided example demonstrates how to calculate the Sobolev norm of a sample function, test a weak solution against the given inputs, and obtain a Schauder estimate related to the regularity of the solution. The results are then printed to the console.

Multiple Choice Questions

1. Which of the following correctly defines the Sobolev space $W^{k,p}(\Omega)$?

    (a) The space of functions with bounded derivatives of order k in $L^p(\Omega)$

    (b) The space of functions in $L^p(\Omega)$ with weak derivatives of order up to k also in $L^p(\Omega)$

    (c) The space of all continuous functions defined on Ω

    (d) The space of functions that are differentiable k times in Ω

2. What is the primary purpose of the Lax-Milgram theorem in the study of elliptic PDEs?

    (a) To provide boundary conditions for solutions of PDEs

    (b) To establish the existence and uniqueness of weak solutions

    (c) To derive numerical methods for solving elliptic equations

    (d) To study the behavior of solutions over time

3. Which statement about the Sobolev embedding theorem is TRUE?

    (a) It guarantees that $W^{k,p}(\Omega)$ is the same space as $L^q(\Omega)$

(b) It relates the smoothness of functions to their integrability properties

(c) It only applies to bounded domains with smooth boundaries

(d) It establishes that all Sobolev spaces are finite-dimensional

4. The Schauder estimates are primarily used to:

    (a) Determine the finiteness of solutions to elliptic equations
    (b) Establish bounds on the regularity of solutions in terms of the coefficients and right-hand side
    (c) Prove the existence of weak solutions for linear systems
    (d) Create numerical algorithms for diverse PDEs

5. What characterizes a weak solution to an elliptic PDE?

    (a) It is a classical solution that satisfies the PDE everywhere in the domain.
    (b) It satisfies the PDE in a distributional sense using test functions.
    (c) It has discontinuous derivatives throughout the domain.
    (d) It approximates the exact solution with finite difference methods.

6. Which of the following is NOT a consequence of the Sobolev embedding theorem?

    (a) The existence of continuous trace functions
    (b) Regularity of solutions to weakly formulated PDEs
    (c) Compactness properties between Sobolev spaces
    (d) Every function in $W^{1,2}(\Omega)$ is continuous

7. The concept of regularity of solutions as discussed in Schauder estimates primarily deals with:

    (a) The temporal behavior of solutions
    (b) The geometric properties of the domain
    (c) The smoothness of the solutions as a function of the input data and coefficients
    (d) The symmetry of solutions regarding specific transformations

Answers:

1. **B: The space of functions in $L^p(\Omega)$ with weak derivatives of order up to k also in $L^p(\Omega)$** This definition captures the essence of Sobolev spaces, where functions must belong to $L^p(\Omega)$ and have weak derivatives up to order k that are also in $L^p(\Omega)$.

2. **B: To establish the existence and uniqueness of weak solutions** The Lax-Milgram theorem provides conditions under which weak solutions exist and are unique for variational formulations of elliptic PDEs, which is essential in their analysis.

3. **B: It relates the smoothness of functions to their integrability properties** The Sobolev embedding theorem explains how the regularity of a function in Sobolev spaces implies certain integrability conditions in other function spaces.

4. **B: Establish bounds on the regularity of solutions in terms of the coefficients and right-hand side** The Schauder estimates give control over the smoothness of solutions based on the smoothness of coefficients in the elliptic PDE and provide critical information about regularity.

5. **B: It satisfies the PDE in a distributional sense using test functions** Weak solutions are defined using integration against test functions, ensuring that even some non-smooth functions can be solutions.

6. **D: Every function in $W^{1,2}(\Omega)$ is continuous** This statement is not universally true; while Sobolev embedding theorems can ensure that functions in certain Sobolev spaces are continuous under specific conditions, it is not a blanket result for all spaces.

7. **C: The smoothness of the solutions as a function of the input data and coefficients** The regularity discussed in Schauder estimates centers around how smooth solutions can be, depending on the smoothness of coefficients and source terms in the elliptic equations.

Chapter 13

The KdV Equation

The Korteweg–de Vries (KdV) equation is a nonlinear partial differential equation that arises in the study of shallow water waves. It was first derived by Korteweg and de Vries in 1895 and has since become a fundamental model in the field of soliton theory. This chapter focuses on the KdV equation, its foundational properties, and various analytical and numerical methods for its solution.

Introduction to Solitons

Solitons are solitary wave solutions that maintain their shape and velocity during propagation. Unlike traditional waves, which disperse and deform over time due to dispersion and nonlinear effects, solitons retain their individuality and can even collide and interact without losing their integrity. Solitons occur in various physical systems, such as water waves, fiber optics, and Bose–Einstein condensates, and their study has led to important insights in nonlinear dynamics and integrable systems.

1 Types of Solitons

There are two main types of solitons: *dispersive solitons* and *modulational solitons*. Dispersive solitons arise from the balance between dispersion and nonlinearity and are typically described by nonlinear Schrödinger equations. Modulational solitons, on the other hand, result from the interplay between dispersion and modulation instability effects and are often described by the KdV equation.

2 The KdV Equation in Physical Context

The KdV equation is a model equation that governs the propagation of weakly nonlinear and weakly dispersive waves in shallow water. It describes the evolution of the wave profile $u(x,t)$ in one spatial dimension. The KdV equation is given by

$$\frac{\partial u}{\partial t} + \alpha u \frac{\partial u}{\partial x} + \beta \frac{\partial^3 u}{\partial x^3} = 0, \tag{13.1}$$

where α and β are constants determining the strength of nonlinearity and dispersion, respectively.

Method of Inverse Scattering Transform

The method of inverse scattering transform (IST) is a powerful technique for solving completely integrable systems, such as the KdV equation. It was developed by Gardner, Greene, Kruskal, and Miura in the 1960s and provides a systematic way to construct soliton solutions through a transformation of the scattering data.

1 Scattering Problem

The scattering problem plays a central role in the IST method. It involves solving a linear ordinary differential equation, known as the *Lax pair*, which is closely related to the KdV equation. The Lax pair associated with the KdV equation (13.1) is given by

$$\frac{\partial \Psi}{\partial x} = \begin{pmatrix} -ik & u \\ -4ik^3 - 2i\alpha & ik \end{pmatrix} \Psi, \tag{13.2}$$

$$\frac{\partial \Psi}{\partial t} = \begin{pmatrix} -6ik^2 & 2iku + \alpha \\ -2ik^3 - i\alpha k - \frac{\beta}{3} & 6ik^2 \end{pmatrix} \Psi, \tag{13.3}$$

where Ψ is a two-component vector function, and k is the spectral parameter.

The scattering problem involves finding a solution Ψ that satisfies appropriate boundary conditions as x tends to $\pm\infty$. The scattering data, defined as the values of Ψ at $x = \pm\infty$, encode the information about the underlying soliton solution.

2 Construction of Solitons

Using the scattering data obtained from the scattering problem, soliton solutions of the KdV equation can be constructed through the inverse scattering transform. The solitons are expressed in terms of the scattering data, and their shapes and velocities are determined by the spectral parameter k.

The soliton solution of the KdV equation takes the form

$$u(x,t) = 2\alpha \frac{\partial^2}{\partial x^2} \log\left[\frac{1+\kappa e^{i\phi}}{1+\kappa e^{-i\phi}}\right], \qquad (13.4)$$

where κ and ϕ are related to the scattering data. The solitons given by (13.4) exhibit remarkable properties, including particle-like behavior, conservation laws, and pairwise elastic interactions.

Numerical Methods

Numerical methods play a crucial role in the study of the KdV equation, enabling the simulation and visualization of soliton dynamics and the exploration of nonlinear phenomena.

1 Finite Difference Methods

Finite difference methods, such as the explicit or implicit Euler method or the Crank-Nicolson method, can be used to discretize the KdV equation in both space and time. These methods approximate the derivatives of the solution using difference quotients and provide a discrete system of equations that can be solved iteratively.

2 Spectral Methods

Spectral methods utilize the Fourier transform or other orthogonal function bases to expand the solution of the KdV equation in a series of basis functions. By truncating the series and applying the Galerkin or collocation method, one obtains a system of ordinary differential equations that can be solved using standard numerical techniques.

3 Pseudospectral Methods

Pseudospectral methods combine the advantages of both finite difference and spectral methods. They approximate the solution and its derivatives using a high-degree polynomial interpolation, typically based on Chebyshev or Legendre nodes. This approach yields accurate results while maintaining computational efficiency.

Analytical Solutions

In addition to numerical methods, the KdV equation possesses a rich set of analytical solutions that provide insight into the behavior of solitons and nonlinear waves.

1 Solitary Wave Solutions

Solitary wave solutions, or solitons, of the KdV equation are obtained through the IST method, as discussed in Section . These solitons are highly localized, maintain their shape and velocity, and exhibit particle-like interactions.

2 Periodic Wave Solutions

The KdV equation also admits periodic wave solutions, known as cnoidal waves. These solutions are periodic in both space and time and have a well-defined amplitude and wavelength. The Jacobi elliptic functions, such as the Jacobi cosine function (cn), play a key role in expressing these solutions.

3 Rational Solutions

Rational solutions of the KdV equation are constructed by taking the ratio of two polynomials in terms of the spectral parameter k. These solutions are expressed in terms of exponential functions and provide a unified framework for describing various soliton and periodic wave solutions.

Applications

The KdV equation and its soliton solutions have far-reaching applications in diverse fields of physics and mathematics. Some notable applications include:

- **Water Waves:** The KdV equation is widely used to model and analyze shallow water waves in rivers, canals, and coastal regions.
- **Fiber Optics:** Soliton solutions of the KdV equation find applications in the transmission of information through optical fibers, where they enable long-distance propagation without distortion.

- **Bose–Einstein Condensates:** The KdV equation provides insights into the behavior of solitary matter wave excitations in Bose–Einstein condensates, leading to better control and manipulation of these quantum systems.

- **Nonlinear Dynamics:** Solitons and the KdV equation contribute to the broader study of nonlinear phenomena and nonlinear dynamical systems, shedding light on the principles of stability, chaos, and pattern formation.

- **Integrable Systems:** The KdV equation and its soliton solutions are part of a larger class of integrable systems, which have deep connections to algebraic geometry, differential geometry, and mathematical physics.

These applications highlight the central role of the KdV equation and its soliton solutions in understanding wave phenomena in various physical and mathematical contexts.

Conclusion

This chapter has provided an overview of the Korteweg–de Vries (KdV) equation, its soliton solutions, and their applications. The KdV equation represents a paradigm in the study of solitons and nonlinear waves, offering insights into the behavior of dispersive and modulational solitons. The inverse scattering transform method serves as a cornerstone for obtaining soliton solutions, while numerical and analytical techniques enable their exploration and analysis. The KdV equation finds applications in fluid dynamics, optics, and quantum mechanics, significantly contributing to our understanding of nonlinear phenomena and integrable systems.

Note: The remaining chapters in this book will explore additional topics in the realm of elliptic partial differential equations, building upon the foundation laid by the KdV equation and its soliton solutions.

Python Code Snippet

Below is a Python code snippet that implements key equations and algorithms discussed in this chapter, specifically focusing on the KdV equation, the method of inverse scattering transform, and numerical methods for solving the equation.

```python
import numpy as np
import matplotlib.pyplot as plt

def kdv_soliton(x, t, alpha, kappa):
    '''
    Calculate the soliton solution of the KdV equation.
    :param x: Spatial variable.
    :param t: Time variable.
    :param alpha: Nonlinearity parameter.
    :param kappa: Amplitude parameter related to soliton speed.
    :return: Soliton solution at (x, t).
    '''
    phi = kappa * (x - 4 * kappa * t)
    return 2 * alpha * (kappa**2) / (1 + kappa**2 * np.cosh(phi) **
      2)

def simulate_kdv_solution(alpha, kappa, x_range, t_values):
    '''
    Simulate and plot soliton solutions of the KdV equation.
    :param alpha: Nonlinearity parameter.
    :param kappa: Amplitude parameter.
    :param x_range: Tuple of spatial domain (min, max).
    :param t_values: List of time values for simulation.
    '''
    x = np.linspace(x_range[0], x_range[1], 400)
    plt.figure(figsize=(10, 6))
    for t in t_values:
        u = kdv_soliton(x, t, alpha, kappa)
        plt.plot(x, u, label=f't={t}')
    plt.title('Soliton Solutions of the KdV Equation')
    plt.xlabel('x')
    plt.ylabel('u(x, t)')
    plt.legend()
    plt.grid()
    plt.show()

def finite_difference_kdv(u0, alpha, beta, dt, dx, t_end):
    '''
    Solve KdV equation using finite difference method.
    :param u0: Initial condition as a numpy array.
    :param alpha: Nonlinearity parameter.
    :param beta: Dispersion parameter.
    :param dt: Time step.
    :param dx: Space step.
    :param t_end: End time for the simulation.
    :return: Time evolution matrix of the solution.
    '''
    rows = int(t_end / dt)
    cols = len(u0)
    u = np.zeros((rows + 1, cols))
    u[0, :] = u0
```

```
        for n in range(0, rows):
            for j in range(1, cols - 1):
                u[n + 1, j] = (u[n, j] - (alpha * u[n, j] * (u[n, j + 1]
                ↪    - u[n, j - 1]) / (2 * dx)) +
                                (beta * (u[n, j + 1] - 2 * u[n, j] + u[n,
                                ↪    j - 1]) / (dx ** 3)) * dt)
        return u

# Parameters and Initial Conditions
alpha = 1.0          # Nonlinearity parameter
beta = -0.2          # Dispersion parameter
x_range = (-20, 20)
t_values = [0, 1, 2]  # Time points to display
kappa = 1.0          # Amplitude parameter for soliton
dx = 0.1             # Spatial step
dt = 0.01            # Time step
t_end = 2            # End time for FD simulation

# Initial condition (Gaussian profile)
x_initial = np.linspace(x_range[0], x_range[1], 400)
u0 = np.exp(-x_initial**2)  # Example initial condition

# Simulations
simulate_kdv_solution(alpha, kappa, x_range, t_values)
u_fd = finite_difference_kdv(u0, alpha, beta, dt, dx, t_end)

# Plotting the results of finite difference method
plt.imshow(u_fd, extent=[x_range[0], x_range[1], 0, t_end],
↪    aspect='auto', cmap='viridis')
plt.colorbar(label='u(x,t)')
plt.title('Finite Difference Method Solution for the KdV Equation')
plt.xlabel('x')
plt.ylabel('t')
plt.show()
```

This code defines several functions:

- **kdv_soliton** computes the soliton solution of the KdV equation based on the given spatial and time variables.
- **simulate_kdv_solution** visualizes the soliton solutions at different time instants.
- **finite_difference_kdv** implements a finite difference method to solve the KdV equation numerically by iterating over time and space.

The provided example showcases both visualizations of solitons through the KdV equation and the numerical approximation using finite differences, reinforcing the key concepts discussed in this chapter.

Multiple Choice Questions

1. What is the primary purpose of the KdV equation in mathematical physics?

    (a) To model turbulent flows

    (b) To describe solitary wave solutions in shallow water

    (c) To predict electromagnetic wave behavior

    (d) To analyze chaotic systems

2. Which of the following types of solitons are associated with the KdV equation?

    (a) Dispersive solitons

    (b) Modulational solitons

    (c) Hyperbolic solitons

    (d) Nonlinear wave solitons

3. The method of inverse scattering transform is primarily used for:

    (a) Solving linear differential equations

    (b) Constructing approximate solutions to complicated PDEs

    (c) Finding soliton solutions of integrable systems

    (d) Analyzing boundary value problems

4. The KdV equation can be expressed mathematically as:

    (a) $\frac{\partial u}{\partial t} + \alpha u^2 + \beta u_{xxxx} = 0$

    (b) $\frac{\partial u}{\partial t} + \alpha u \frac{\partial u}{\partial x} + \beta u_{xx} = 0$

    (c) $\frac{\partial u}{\partial t} + \alpha u \frac{\partial u}{\partial x} + \beta \frac{\partial^3 u}{\partial x^3} = 0$

    (d) $\frac{\partial u}{\partial t} - \alpha \frac{\partial^2 u}{\partial x^2} = 0$

5. In the context of the KdV equation, what role does the spectral parameter k play?

    (a) It determines the amplitude of the solution

    (b) It represents the wave number

    (c) It governs the dispersion relation

    (d) It influences the nonlinear interactions

6. Which numerical method is typically employed to approximate solutions of the KdV equation?

    (a) Finite Element Method
    (b) Finite Difference Method
    (c) Polynomial Approximation Method
    (d) Perturbation Method

7. Cnoidal waves, which are periodic solutions of the KdV equation, are expressed in terms of which mathematical functions?

    (a) Bessel Functions
    (b) Airy Functions
    (c) Jacobi Elliptic Functions
    (d) Hypergeometric Functions

Answers:

1. **B: To describe solitary wave solutions in shallow water** The KdV equation is a fundamental model for studying weakly nonlinear and dispersive waves in shallow water, highlighting its role in describing solitary wave solutions.

2. **B: Modulational solitons** Modulational solitons are specifically associated with the KdV equation, arising from the balance between dispersion and the nonlinear effects in wave propagation.

3. **C: Finding soliton solutions of integrable systems** The method of inverse scattering transform is a technique developed to systematically construct soliton solutions for integrable systems, including the KdV equation.

4. **C:** $\frac{\partial u}{\partial t} + \alpha u \frac{\partial u}{\partial x} + \beta \frac{\partial^3 u}{\partial x^3} = 0$ This equation represents the standard form of the KdV equation, illustrating the interplay between nonlinearity (first term) and dispersion (third term).

5. **B: It represents the wave number** In the context of the KdV equation, the spectral parameter k is associated with the wave number and it governs the properties of the soliton solutions.

6. **B: Finite Difference Method** The finite difference method is commonly used for numerically solving the KdV equation, as it discretizes both space and time to approximate the solution iteratively.

7. **C: Jacobi Elliptic Functions** Cnoidal waves, which provide periodic wave solutions to the KdV equation, are expressed in terms of Jacobi elliptic functions, enabling characterization of wavelike phenomena.

Chapter 14

The Ricci Flow Equation

Geometric Analysis Background

Geometric analysis is a branch of mathematics that combines techniques from differential geometry and partial differential equations to study geometric structures. In this chapter, we focus on the Ricci flow equation, a powerful tool in geometric analysis for understanding the geometry of manifolds.

Hamilton's Formulation

The Ricci flow equation was first introduced by Richard S. Hamilton in 1982 as a nonlinear parabolic partial differential equation. It is defined on a Riemannian manifold and describes how the metric on the manifold evolves under the flow. The equation is given by:

$$\frac{\partial g_{\alpha\beta}}{\partial t} = -2R_{\alpha\beta}, \qquad (14.1)$$

where $g_{\alpha\beta}$ is the metric tensor of the manifold, $R_{\alpha\beta}$ is its Ricci curvature tensor, and t represents the flow parameter.

Long-Time Behavior

One of the key questions in studying the Ricci flow equation is understanding its long-time behavior. Hamilton showed that under certain conditions, the Ricci flow exists for all time and converges to a metric with constant curvature. This is known as the Hamilton-Ivey pinching theorem and has important implications in geometric analysis.

Numerical Simulations

Solving the Ricci flow equation analytically for general manifolds is often challenging. Therefore, numerical simulations play a crucial role in understanding the behavior of the flow. Several numerical methods have been developed, including finite difference schemes, finite element methods, and parabolic mesh generation techniques.

Applications in Geometry and Topology

The Ricci flow equation has wide-ranging applications in both geometry and topology. By studying the evolution of the metric along the flow, researchers have made significant progress in areas such as the classification of manifolds, the understanding of geometric singularities, and the proof of the Poincaré conjecture in three dimensions.

Note: The Ricci flow equation has connections to other important topics in mathematics, such as the Ricci flow on non-compact manifolds, the Ricci flow on singular spaces, and the relationship between the Ricci flow and other geometric evolution equations. These topics will be explored in further detail in future chapters.

Python Code Snippet

Below is a Python code snippet that implements the Ricci flow equation on a 2D manifold using finite difference methods for numerical simulation. The code initializes a Riemannian metric and evolves it according to the Ricci flow, visualizing the metric's evolution at various time steps.

```python
import numpy as np
import matplotlib.pyplot as plt

def initialize_metric(grid_size):
    '''
    Initialize the metric tensor for the 2D manifold.
    :param grid_size: Size of the grid for the manifold.
    :return: Initialized metric tensor (2D array).
    '''
    return np.ones((grid_size, grid_size))

def compute_curvature(metric):
    '''
    Compute the Ricci curvature tensor from the metric tensor.
    :param metric: Metric tensor (2D array).
    :return: Ricci curvature (2D array).
    '''
    # Placeholder code for curvature computation
    curvature = np.zeros_like(metric)  # Simplified for
    ↪ demonstration
    for i in range(1, metric.shape[0] - 1):
        for j in range(1, metric.shape[1] - 1):
            curvature[i, j] = metric[i, j]  # Simplistic
            ↪ representation
    return curvature

def ricci_flow(metric, time_step, num_steps):
    '''
    Evolve the metric tensor according to the Ricci flow.
    :param metric: Initial metric tensor (2D array).
    :param time_step: Time step for the numerical scheme.
    :param num_steps: Number of time evolution steps.
    :return: List of evolved metric tensors.
    '''
    metrics = [metric.copy()]
    for _ in range(num_steps):
        curvature = compute_curvature(metrics[-1])
        # Update the metric based on the Ricci flow equation
        new_metric = metrics[-1] - 2 * curvature * time_step
        metrics.append(new_metric)
    return metrics

def visualize_metrics(metrics):
    '''
    Visualize the progression of the metrics over time.
    :param metrics: List of metric tensors (2D arrays).
    '''
    fig, axes = plt.subplots(1, len(metrics), figsize=(15, 5))
    for ax, metric in zip(axes, metrics):
        ax.imshow(metric, cmap='hot', interpolation='nearest')
        ax.axis('off')
    plt.show()
```

```
# Parameters
grid_size = 100    # Size of the manifold grid
time_step = 0.01   # Time step for simulation
num_steps = 10     # Number of evolution steps

# Initialize the metric tensor
initial_metric = initialize_metric(grid_size)

# Evolve the metric according to Ricci flow
metrics = ricci_flow(initial_metric, time_step, num_steps)

# Visualize the evolution of the metric
visualize_metrics(metrics)
```

This code defines several functions:

- initialize_metric initializes the metric tensor for a 2D manifold as a uniform metric.
- compute_curvature computes a placeholder Ricci curvature tensor from the metric tensor.
- ricci_flow evolves the metric tensor over time according to the Ricci flow equation, updating the metric using the computed curvature.
- visualize_metrics visualizes the evolution of the metric tensors at different time steps.

The provided example initializes a metric on a 2D grid, then simulates the Ricci flow over a specified number of time steps, and visualizes the metric's evolution.

Multiple Choice Questions

1. Who first introduced the Ricci flow equation?

    (a) Andrew Wiles

    (b) John Nash

    (c) Richard S. Hamilton

    (d) Henri Poincaré

2. The Ricci flow equation describes the evolution of which of the following?

    (a) Geodesics on a manifold

    (b) The metric tensor of a Riemannian manifold

(c) The curvature scalar of a manifold

(d) The volume of a manifold

3. The key result known as the Hamilton-Ivey pinching theorem pertains to:

    (a) The convergence of metrics with minimal Ricci curvature

    (b) The long-term behavior of the Ricci flow on compact manifolds

    (c) The applicability of Ricci flow to algebraic varieties

    (d) The stability of the Einstein equations

4. Which numerical method is NOT commonly used to solve the Ricci flow equation?

    (a) Finite difference schemes

    (b) Finite element methods

    (c) Spectral methods

    (d) Monte Carlo simulations

5. The study of the Ricci flow contributes significantly to our understanding of which mathematical area?

    (a) Algebraic topology

    (b) Functional analysis

    (c) Differential geometry

    (d) Number theory

6. What is one of the primary motivations for studying the Ricci flow equation?

    (a) To classify Riemann surfaces

    (b) To prove the Poincaré conjecture

    (c) To study the behavior of heat equations

    (d) To analyze financial models

7. True or False: The Ricci flow can be applied to non-compact manifolds.

    (a) True

(b) False

Answers:

1. **C: Richard S. Hamilton** Richard S. Hamilton first introduced the Ricci flow equation in 1982. His work laid the groundwork for significant advancements in geometric analysis.

2. **B: The metric tensor of a Riemannian manifold** The Ricci flow describes how the metric tensor on a Riemannian manifold evolves over time, driven by the curvature of the manifold.

3. **B: The long-term behavior of the Ricci flow on compact manifolds** The Hamilton-Ivey pinching theorem establishes conditions under which the Ricci flow exists for all time and converges to a metric of constant curvature on compact manifolds.

4. **D: Monte Carlo simulations** While finite difference schemes, finite element methods, and spectral methods are common numerical approaches, Monte Carlo simulations are not typically associated with solving the Ricci flow equation.

5. **C: Differential geometry** The study of the Ricci flow is fundamentally rooted in differential geometry, where it is used to understand and classify geometric structures.

6. **B: To prove the Poincaré conjecture** One of the major motivations for studying the Ricci flow was its application in proving the Poincaré conjecture for three-dimensional manifolds, a landmark result in topology.

7. **A: True** The Ricci flow can indeed be applied to non-compact manifolds. Its extensions to non-compact settings are an area of active research and lead to a deeper understanding of geometric evolution.

Chapter 15

Variational Methods in Elliptic PDEs

Calculus of Variations

The calculus of variations is a branch of mathematics that deals with finding the extremum of a functional. In the context of elliptic partial differential equations (PDEs), variational methods play a crucial role in solving problems involving these equations. By representing the PDE as a variational problem, we can find solutions by minimizing or maximizing a functional over a suitable function space.

Euler-Lagrange Equations

The Euler-Lagrange equations form the necessary conditions for a function to be an extremum of a functional. In the context of elliptic PDEs, these equations correspond to the variational formulation of the PDE. By deriving and solving the Euler-Lagrange equations associated with a given functional, we can obtain solutions to the corresponding elliptic PDE.

Functional Spaces

To apply variational methods to elliptic PDEs, we need to carefully define the function spaces over which our variational problem is formulated. Common function spaces used in this context include Sobolev spaces, which provide the appropriate setting for weak solutions of elliptic PDEs. These function spaces ensure that the functions satisfy the necessary regularity conditions for the solutions of the PDE.

Minimization Techniques

Once the variational problem is formulated and the function space is defined, we can employ various minimization techniques to find the extremum of the functional. This may involve utilizing optimization algorithms or applying calculus tools such as the method of Lagrange multipliers. The choice of minimization technique depends on the nature of the problem and the functional involved.

Applications in Physics and Engineering

Variational methods in elliptic PDEs find a wide range of applications in physics and engineering. They are commonly used in problems related to potential theory, where the Laplace equation plays a fundamental role. Variational methods also arise in the study of elasticity, fluid mechanics, electromagnetism, and other areas where elliptic PDEs are essential for describing physical phenomena.

Note: Variational methods have connections to other areas of mathematics, such as optimal control theory, functional analysis, and geometric measure theory. These connections provide further insight into the deep interplay between variational methods and elliptic PDEs, but exploring them is beyond the scope of this chapter.

Python Code Snippet

Below is a Python code snippet that implements variational methods for solving elliptic partial differential equations using the finite element method. Specifically, this code demonstrates how to solve the Poisson equation on a unit square domain.

```
import numpy as np
import matplotlib.pyplot as plt
from scipy.sparse import csr_matrix
```

```python
from scipy.sparse.linalg import spsolve

def generate_mesh(n):
    '''
    Generate a uniform mesh for the unit square [0, 1] x [0, 1].
    :param n: Number of divisions on each axis (n x n grid).
    :return: The mesh grid points and the corresponding elements.
    '''
    x = np.linspace(0, 1, n+1)  # x coordinates
    y = np.linspace(0, 1, n+1)  # y coordinates
    points = np.array(np.meshgrid(x, y)).T.reshape(-1, 2)
    elements = []

    for i in range(n):
        for j in range(n):
            elements.append([i*(n+1) + j, (i+1)*(n+1) + j,
                (i+1)*(n+1) + (j+1)])  # Lower triangle
            elements.append([i*(n+1) + j, (i+1)*(n+1) + (j+1),
                i*(n+1) + (j+1)])  # Upper triangle

    return points, np.array(elements)

def assemble_system(points, elements):
    '''
    Assemble the stiffness matrix and load vector for the finite
      element formulation.
    :param points: The mesh points.
    :param elements: The connectivity of the mesh.
    :return: Stiffness matrix A and load vector b.
    '''
    n_points = len(points)
    A = np.zeros((n_points, n_points))
    b = np.zeros(n_points)

    for el in elements:
        # Extract the local nodes
        pts = points[el]
        area = 0.5 * np.abs(np.linalg.det(np.array(
            [[1, pts[0][0], pts[0][1]],
             [1, pts[1][0], pts[1][1]],
             [1, pts[2][0], pts[2][1]]])))

        # Contribution to the global stiffness matrix
        for i in range(3):
            for j in range(3):
                A[el[i], el[j]] += area * np.linalg.inv(np.array(
                    [[1, pts[0][0], pts[0][1]],
                     [1, pts[1][0], pts[1][1]],
                     [1, pts[2][0], pts[2][1]]]))[i, j]

        # Load vector assuming a unit load
        for i in range(3):
            b[el[i]] += area / 3  # Average load
```

```
        return csr_matrix(A), b

def solve_poisson(n):
    '''
    Solve the Poisson equation using the finite element method on a
    ↪ unit square.
    :param n: Number of divisions for the mesh.
    :return: Solution vector u.
    '''
    points, elements = generate_mesh(n)
    A, b = assemble_system(points, elements)

    # Apply Dirichlet boundary conditions (u = 0 on the boundary)
    for i in range(len(points)):
        if points[i, 0] == 0 or points[i, 0] == 1 or points[i, 1] ==
        ↪ 0 or points[i, 1] == 1:
            A[i, :] = 0
            A[i, i] = 1
            b[i] = 0

    # Solve the linear system
    u = spsolve(A, b)
    return points, u

# Parameters for the problem
n = 20  # Number of subdivisions
points, solution = solve_poisson(n)

# Plotting the solution
plt.tripcolor(points[:, 0], points[:, 1], solution, shading='flat',
↪ cmap='viridis')
plt.colorbar(label='u(x, y)')
plt.title('Solution to the Poisson Equation')
plt.xlabel('x')
plt.ylabel('y')
plt.show()
```

This code defines several functions:

- generate_mesh generates a uniform mesh for the unit square [0, 1] x [0, 1] for a specified number of subdivisions.
- assemble_system assembles the global stiffness matrix and load vector for the finite element method based on the mesh.
- solve_poisson implements the finite element method to solve the Poisson equation by applying boundary conditions and solving the resulting linear system.

The final part of the code visualizes the solution to the Poisson equation on a unit square using a trip-color plot. This effectively demonstrates the application of variational methods in elliptic PDE solving.

Multiple Choice Questions

1. In the context of the calculus of variations, a functional is considered to be:

    (a) A function of several variables

    (b) A mapping from a function space to the real numbers

    (c) A differential equation

    (d) An operator on vector spaces

2. The Euler-Lagrange equations provide:

    (a) Necessary conditions for extremality of a functional

    (b) Sufficient conditions for a function to be continuous

    (c) Conditions for convergence of sequences in functional spaces

    (d) Solutions to nonlinear partial differential equations

3. Which of the following function spaces is most commonly used in variational methods for elliptic PDEs?

    (a) $L^p spaces Sobolev spaces$

(b) Banach spaces

(c) Hilbert spaces

4. When applying minimization techniques in variational problems, the method of Lagrange multipliers is used to:

    (a) Solve linear equations

    (b) Handle constraints in the optimization problem

    (c) Maximize the functional without constraints

    (d) Classify the type of differential equation

5. Which application area does NOT typically utilize variational methods with elliptic PDEs?

    (a) Electromagnetism

    (b) Quantum mechanics

(c) Potential theory

(d) Artificial intelligence

6. The Sobolev space $H^1(\Omega)$ consists of functions that are:

    (a) Square integrable

    (b) Absolutely integrable

    (c) Integrable along with their first derivatives

    (d) Continuous on the domain

7. Variational methods allow us to transform a PDE problem into:

    (a) A system of ordinary differential equations

    (b) An optimization problem involving a functional

    (c) A linear programming problem

    (d) A system of algebraic equations

Answers:

1. **B: A mapping from a function space to the real numbers** A functional takes functions as inputs and produces a real number, serving as a foundational concept in the calculus of variations.

2. **A: Necessary conditions for extremality of a functional** The Euler-Lagrange equations derive from the principle of least action and provide necessary conditions for a function to be an extremum of a given functional.

3. **B: Sobolev spaces** Sobolev spaces, such as $W^{k,p}$, incorporate both function regularity and integrability conditions suitable for weak solutions of elliptic PDEs.

4. **B: Handle constraints in the optimization problem** The method of Lagrange multipliers is employed to optimize a function subject to constraints, thereby facilitating constrained optimization problems.

5. **D: Artificial intelligence** While variational methods play significant roles in fields like electromagnetism, quantum mechanics, and potential theory, they are not predominantly associated with artificial intelligence.

6. **C: Integrable along with their first derivatives** The Sobolev space $H^1(\Omega)$ consists of functions that are square integrable and have square integrable weak derivatives, ensuring appropriate regularity.

7. B: An optimization problem involving a functional

Variational methods reformulate differential equations into optimization problems, enabling the search for extremum values of functionals that correspond to solutions of the original PDEs.

Chapter 16

The Heat Equation on Manifolds

Differential Geometry Basics

In this chapter, we delve into the study of the heat equation on manifolds, which are generalizations of smooth surfaces to higher dimensions. To understand this topic, it is crucial to have a solid foundation in **differential geometry**. Differential geometry focuses on the properties of smooth manifolds, such as their tangent spaces, metrics, and connections. These concepts provide the tools necessary to define and analyze the heat equation on manifolds.

Formulation on Riemannian Manifolds

In the study of the heat equation on manifolds, we typically work with **Riemannian manifolds**, which are smooth manifolds equipped with a Riemannian metric. The Riemannian metric defines a notion of distance and inner product on the tangent spaces of the manifold at each point. We introduce the Laplace-Beltrami operator, a differential operator that generalizes the Laplacian to manifolds, as it plays a central role in the formulation of the heat equation.

Spectral Properties

Understanding the spectral properties of the Laplace-Beltrami operator is key to analyzing the heat equation on manifolds. We explore spectral theory, which provides insights into the eigenvalues and eigenfunctions of the Laplace-Beltrami operator. The spectral properties elucidate the behavior of the heat equation, allowing us to investigate questions of stability, convergence, and long-time behavior.

Numerical Approaches

To numerically solve the heat equation on manifolds, we require discretization methods tailored to the manifold's structure. Finite element methods and finite difference methods adapted to manifolds have been developed to approximate solutions to the heat equation. These numerical approaches enable us to study the behavior of the heat equation on manifolds when an analytical solution is not available or difficult to obtain.

Applications in Geometry

The study of the heat equation on manifolds has far-reaching applications in geometry. It provides insights into geometric properties such as curvature, volume, and geodesics. By examining the behavior of the heat equation on manifolds, we can gain a deeper understanding of geometric structures and explore important geometric phenomena, such as Ricci flow and curvature evolution.

Within the realm of mathematical physics, the heat equation on manifolds also plays a vital role. It connects to various physical phenomena, including diffusion processes, heat conduction, and stochastic processes on curved spaces. By studying the heat equation on manifolds, we can uncover profound connections between mathematical and physical concepts.

The heat equation on manifolds represents a rich field of research, combining concepts from differential geometry, spectral theory, and numerical analysis. Through the exploration of this chapter, we aim to develop a solid understanding of the heat equation on manifolds and its applications in geometry and mathematical physics.Certainly! Below is a comprehensive Python code snippet relevant to the heat equation on manifolds, properly formatted with LaTeX and the 'minted' package, as requested.

"'latex

Python Code Snippet

Below is a Python code snippet that implements the numerical solution to the heat equation on Riemannian manifolds using the finite difference method.

```python
import numpy as np

def laplace_beltrami_operator(mesh, diffusion_coefficient):
    '''
    Compute the Laplace-Beltrami operator on a given mesh
    (discretized manifold).
    :param mesh: List of vertices and corresponding connections
    (edges) of the manifold.
    :param diffusion_coefficient: Coefficient for diffusion in the
    heat equation.
    :return: The Laplace-Beltrami operator matrix.
    '''
    n = len(mesh['vertices'])
    L = np.zeros((n, n))  # Initialize operator matrix

    for i, vertex in enumerate(mesh['vertices']):
        for j in mesh['adjacencies'][i]:
            weight = diffusion_coefficient / np.linalg.norm(vertex -
                mesh['vertices'][j])
            L[i, j] = -weight
            L[i, i] += weight  # Diagonal entry is sum of adjacent
                weights

    return L

def heat_equation_on_manifold(mesh, initial_temperature, time_steps,
    dt):
    '''
    Solve the heat equation on the manifold using a semi-discrete
    method.
    :param mesh: The mesh representing the manifold.
    :param initial_temperature: Initial temperature distribution as
    a numpy array.
    :param time_steps: Number of time steps for the simulation.
    :param dt: Time step size.
    :return: Temperature distribution over time.
    '''
    L = laplace_beltrami_operator(mesh, diffusion_coefficient=1.0)
        # Assuming diffusion coefficient of 1.0
```

```python
    temperature_distribution = np.zeros((time_steps,
        len(initial_temperature)))
    temperature_distribution[0, :] = initial_temperature

    for t in range(1, time_steps):
        temperature_distribution[t, :] =
            temperature_distribution[t-1, :] + dt * (L @
            temperature_distribution[t-1, :])

    return temperature_distribution

# Example usage
if __name__ == "__main__":
    # Define a simple mesh (2D grid as an example)
    mesh = {
        'vertices': np.array([[0, 0], [1, 0], [0, 1], [1, 1]]),
        'adjacencies': [[1, 2], [0, 3], [0, 3], [1, 2]],
    }
    initial_temperature = np.array([100, 0, 0, 0])  # Initial
        temperatures at each vertex
    time_steps = 50  # Total time steps
    dt = 0.01  # Time step size

    # Solve the heat equation on the defined manifold
    temperature_evolution = heat_equation_on_manifold(mesh,
        initial_temperature, time_steps, dt)

    # Output the final temperature distribution
    print("Final Temperature Distribution:",
        temperature_evolution[-1])
```

This code defines the following functions:

- `laplace_beltrami_operator` computes the Laplace-Beltrami operator based on the provided mesh representing the manifold.
- `heat_equation_on_manifold` implements the finite difference method to solve the heat equation over the discretized manifold for a specified number of time steps.

The example usage initializes a simple 2D mesh and calculates the temperature evolution according to the heat equation, printing the final temperature distribution after simulation.

Multiple Choice Questions

1. What is the primary purpose of the Laplace-Beltrami operator in the context of the heat equation on Riemannian manifolds?

(a) To calculate the curvature of the manifold

   (b) To extend the concept of the Laplacian to higher-dimensional manifolds

   (c) To define geodesics on the manifold

   (d) To find eigenvalues of the manifold

2. Which of the following is NOT a property of Riemannian manifolds?

   (a) They have an inner product defined on tangent spaces

   (b) They can be considered as metric spaces

   (c) They must have a positive curvature

   (d) They possess a smooth and continuous structure

3. Which numerical method is commonly employed for solving the heat equation on manifolds?

   (a) Finite Element Method

   (b) Runge-Kutta Method

   (c) Monte Carlo Method

   (d) Gradient Descent

4. In spectral theory concerning the Laplace-Beltrami operator, what do we investigate?

   (a) The topology of the manifold

   (b) The eigenvalues and eigenfunctions of the operator

   (c) The geodesic completeness of the manifold

   (d) The integrability condition of the manifold

5. The heat equation on manifolds is closely related to which of the following physical phenomena?

   (a) Wave propagation

   (b) Fluid dynamics

   (c) Diffusion processes

   (d) Electromagnetism

6. The heat equation on Riemannian manifolds can provide insights into which of the following geometric properties?

(a) The degree of differentiability of the manifold

(b) The curvatures of the manifold

(c) The topology of the manifold

(d) The compactness of the manifold

7. True or False: The heat equation on manifolds can be understood as a diffusion equation describing how heat spreads over the manifold.

(a) True

(b) False

Answers:

1. **B: To extend the concept of the Laplacian to higher-dimensional manifolds** The Laplace-Beltrami operator generalizes the traditional Laplacian in Euclidean spaces to Riemannian manifolds, which is essential for formulating the heat equation within this context.

2. **C: They must have a positive curvature** Riemannian manifolds can have various types of curvature (positive, negative, or zero), and thus they do not necessarily need to have positive curvature.

3. **A: Finite Element Method** The Finite Element Method (FEM) is widely used for numerically solving partial differential equations like the heat equation on manifolds, as it can effectively handle the geometric complexity.

4. **B: The eigenvalues and eigenfunctions of the operator** Spectral theory focuses on the properties of the eigenvalues and eigenfunctions of the Laplace-Beltrami operator, which helps in analyzing the behavior of the heat equation.

5. **C: Diffusion processes** The heat equation models diffusion processes as it describes how heat (or other quantities) spreads over a given space—one of the fundamental phenomena modeled by this equation.

6. **B: The curvatures of the manifold** The behavior of the heat equation can reveal information about the curvature of the manifold due to the interplay between heat diffusion and geometric properties.

7. **A: True** The heat equation on manifolds can indeed be seen as a diffusion equation that describes how heat propagates through the manifold, confirming its nature in both mathematical and physical frameworks.

Chapter 17

The Allen-Cahn Equation

Phase Field Models

Phase field models provide a valuable mathematical framework for studying phase transitions and interface dynamics. In the context of materials science and physics, phase transitions occur when a system undergoes a change in its thermodynamic state, resulting in the formation or disappearance of distinct phases. The Allen-Cahn equation, named after John W. Cahn and John E. Allen, is a well-known phase field model that describes the evolution of interfaces between phases in a system.

Analytical Solutions

Solving the Allen-Cahn equation analytically is a challenging task due to its nonlinearity. However, in some special cases, exact solutions can be obtained. These solutions provide valuable insights into the behavior of the interface. One such example is the planar front solution, which describes a flat interface moving at a constant speed. Understanding these analytical solutions aids in developing intuition and serves as a benchmark for numerical methods.

Numerical Simulations

Given the difficulty of obtaining analytical solutions for the Allen-Cahn equation, numerical simulations play a vital role in studying its behavior. Various numerical methods have been developed to approximate the solution of the Allen-Cahn equation, including finite difference methods, finite element methods, and spectral methods. These techniques discretize the equation in space and time and provide numerical solutions that capture the evolution of interfaces between phases.

Application to Material Science

The Allen-Cahn equation has significant applications in material science, particularly in the study of microstructure evolution. It provides insights into the formation and growth of various microstructures, such as grains, domains, and interfaces, in materials. By simulating the Allen-Cahn equation, researchers can investigate the relationship between material properties, interface dynamics, and the resulting microstructure. This understanding is crucial for materials design and optimization.

Pattern Formation and Dynamics

In addition to its applications in material science, the Allen-Cahn equation is widely used to study pattern formation and dynamical systems. The equation's solutions exhibit rich behavior, including the formation of intricate patterns and the emergence of complex dynamics. By exploring the numerical and analytical solutions of the Allen-Cahn equation, researchers gain valuable insights into pattern formation mechanisms and the behavior of nonlinear systems.

The study of the Allen-Cahn equation provides a deep understanding of phase transitions, interface dynamics, and pattern formation. Its applications span across various disciplines, including material science, physics, and mathematics itself. This chapter focuses on the mathematical aspects of the Allen-Cahn equation, presenting analytical and numerical techniques for solving and analyzing its solutions. By delving into the intricacies of this equation, we can uncover fundamental principles of phase transitions and gain insights that extend beyond this specific model."'latex

Python Code Snippet

Below is a Python code snippet that implements the Allen-Cahn equation using numerical methods for simulating phase field dynamics.

```python
import numpy as np
import matplotlib.pyplot as plt

def allen_cahn(phi, alpha, dt, dx, iterations):
    '''
    Solve the Allen-Cahn equation using a simple finite difference
    ↪ scheme.
    :param phi: Initial condition for the phase field.
    :param alpha: Parameter that controls the evolution.
    :param dt: Time step for the simulation.
    :param dx: Spatial step for the simulation.
    :param iterations: Number of iterations to run the simulation.
    :return: Array containing the phase field after specified
    ↪ iterations.
    '''
    for _ in range(iterations):
        laplacian_phi = (np.roll(phi, 1) + np.roll(phi, -1) - 2 *
        ↪ phi) / dx**2
        phi += dt * (alpha * laplacian_phi + phi * (1 - phi) * (phi
        ↪ - 0.5))
    return phi

def initialize_phi(N):
    '''
    Initialize the phase field variable.
    :param N: Number of grid points.
    :return: Array containing the initialized phase field.
    '''
    phi = np.random.rand(N)
    phi[N//2 - 10:N//2 + 10] = 1  # Create initial phase
    return phi

# Simulation parameters
N = 100       # Number of grid points
dx = 1.0      # Spatial step
dt = 0.01     # Time step
iterations = 1000  # Number of iterations
alpha = 1.0   # Parameter for Allen-Cahn equation

# Initialize the phase field
phi = initialize_phi(N)
```

```
# Store the results for plotting
results = []
for _ in range(10):
    phi = allen_cahn(phi, alpha, dt, dx, iterations // 10)
    results.append(phi.copy())

# Plotting the results
plt.figure(figsize=(10, 6))
for i in range(len(results)):
    plt.plot(results[i], label=f'Step {i+1}')
plt.title('Evolution of the Phase Field via Allen-Cahn Equation')
plt.xlabel('Spatial grid points')
plt.ylabel('Phase field $\phi$')
plt.legend()
plt.show()
```

This code defines two functions:

- `allen_cahn` implements a finite difference scheme to solve the Allen-Cahn equation iteratively.
- `initialize_phi` initializes the phase field with a random distribution and a defined initial condition.

The provided example runs a simulation of the Allen-Cahn equation, visualizing the evolution of the phase field over time, illustrating how interfaces develop and change under the influence of the underlying dynamics. "'

Multiple Choice Questions

1. What is the primary physical phenomenon modeled by the Allen-Cahn equation?

    (a) Heat conduction

    (b) Diffusion of particles

    (c) Phase transitions and interface dynamics

    (d) Fluid flow

2. Which of the following statements describes the unique feature of the Allen-Cahn equation?

    (a) It is a linear partial differential equation.

    (b) It incorporates a reaction-diffusion type dynamic.

    (c) It only describes equilibrium states.

(d) It cannot exhibit non-linear behavior.

3. In the context of numerical methods for the Allen-Cahn equation, which approach is commonly employed to capture the evolution of interfaces?

   (a) Laplace transforms

   (b) Spectral methods

   (c) Finite element methods

   (d) Monte Carlo simulations

4. Which type of solution does not typically arise from the Allen-Cahn equation?

   (a) Planar front solutions

   (b) Coinciding phase interfaces

   (c) Complex multi-phase patterns

   (d) Steady-state microstructures

5. The Allen-Cahn equation is primarily used in which of the following scientific fields?

   (a) Quantum physics and mechanics

   (b) Epidemiology and population dynamics

   (c) Material science and microstructure evolution

   (d) Telecommunications and signal processing

6. What is an important aspect of interface dynamics described by the Allen-Cahn equation?

   (a) Interfaces cannot change over time.

   (b) Interfaces evolve only due to external forces.

   (c) Interfaces exhibit velocity-dependent motion.

   (d) Interfaces are only characterized by their widths.

7. True or False: The Allen-Cahn equation has been shown to exhibit soliton-like solutions under certain conditions.

   (a) True

   (b) False

Answers:

1. **C: Phase transitions and interface dynamics** The Allen-Cahn equation is specifically designed to model the evolution of phase interfaces, which is a critical aspect in studying materials undergoing phase transitions.

2. **B: It incorporates a reaction-diffusion type dynamic** The Allen-Cahn equation is nonlinear and describes how the concentration of different phases in a material evolves over time due to diffusion and reactions.

3. **C: Finite element methods** Finite element methods are a key numerical approach used to solve the Allen-Cahn equation, as they allow for the effective discretization of complex geometries and interfaces.

4. **B: Coinciding phase interfaces** While the Allen-Cahn equation models various interfaces and their dynamics, coinciding phase interfaces (i.e., interfaces that do not change) are not typical solutions as they imply a lack of evolution.

5. **C: Material science and microstructure evolution** The primary application of the Allen-Cahn equation is in material science, where it helps understand the formation of microstructures during phase transitions.

6. **C: Interfaces exhibit velocity-dependent motion** The dynamics of interfaces described by the Allen-Cahn equation reflect how the speed of interface evolution can depend on factors such as curvature and gradients in phase concentration.

7. **A: True** Under certain conditions, the Allen-Cahn equation can exhibit soliton-like solutions, which are stable waveforms that maintain their shape while propagating, a characteristic of some nonlinear dynamical systems.

Chapter 18

The Ginzburg-Landau Equations

Superconductivity Theories

Superconductivity is a fascinating phenomenon in condensed matter physics where certain materials exhibit zero electrical resistance and expel magnetic fields below a critical temperature. The Ginzburg-Landau equations, named after Vitaly Ginzburg and Lev Landau, provide a mathematical framework for describing superconductivity phenomena. These equations are derived from considering the behavior of the complex-valued order parameter, which characterizes the superconducting state.

Mathematical Formulation

The Ginzburg-Landau equations describe the evolution of the order parameter in superconducting materials. The equations can be expressed as follows:

$$\left(-i\hbar\nabla - \frac{2e}{c}\mathbf{A}(\mathbf{x})\right)\Psi(\mathbf{x}) = \alpha\Psi(\mathbf{x}) + \beta|\Psi(\mathbf{x})|^2\Psi(\mathbf{x})$$

$$\nabla \cdot \mathbf{B}(\mathbf{x}) = 0$$

$$\nabla \times \mathbf{B}(\mathbf{x}) = \frac{4\pi}{c}\mathbf{J}(\mathbf{x}) + \frac{4\pi e}{c}\text{Im}\left(\Psi(\mathbf{x})\nabla\Psi^*(\mathbf{x})\right),$$

where $\Psi(\mathbf{x})$ is the order parameter, $\mathbf{A}(\mathbf{x})$ is the vector potential, $\mathbf{B}(\mathbf{x})$ is the magnetic field, $\mathbf{J}(\mathbf{x})$ represents the current density, α and β are material-specific parameters, $\hbar$ is the reduced Planck constant, e is the elementary charge, and c denotes the speed of light.

Vortex Solutions

Vortices are topological defects in the superconducting order parameter field. They are characterized by a non-zero winding number and possess quantized magnetic flux. The Ginzburg-Landau equations describe the presence and behavior of these vortices. Numerical techniques, such as finite difference methods and spectral methods, are commonly used to solve the equations and obtain vortex solutions. These solutions provide insights into the stability, interactions, and dynamics of vortices in superconductors.

Numerical Techniques

Solving the Ginzburg-Landau equations requires the use of numerical methods due to the complexity and nonlinearity of the equations. Finite difference methods, such as the Crank-Nicolson method or the alternating direction implicit (ADI) method, discretize the equations in space and time and solve them iteratively. Spectral methods, such as Fourier or Chebyshev spectral methods, exploit the inherent periodicity of the problem to approximate the solution with spectral accuracy. These numerical techniques offer efficient and accurate approaches to studying the Ginzburg-Landau equations.

Applications in Condensed Matter Physics

The Ginzburg-Landau equations have significant applications in condensed matter physics, particularly in the study of superconductivity. By simulating and analyzing solutions of these equations, researchers can investigate various phenomena, such as the formation and behavior of vortices, the critical temperature for superconductivity, and the response of superconducting materials to external magnetic fields. This understanding is vital for designing and optimizing superconductive devices with enhanced performance and efficiency.

The study of the Ginzburg-Landau equations provides valuable insights into the fundamental principles of superconductivity. Through analytical and numerical investigations, researchers can deepen their understanding of the behavior of order parameters, magnetic fields, and current densities in superconducting materials. This chapter has presented the mathematical formulation of the Ginzburg-Landau equations, discussed vortex solutions and numerical techniques for solving the equations, and highlighted their applications in condensed matter physics. Further exploration of these equations will provide a deeper understanding of superconductivity and its potential future applications.

Python Code Snippet

Below is a Python code snippet that implements the Ginzburg-Landau equations for a superconducting system. The code sets up the necessary parameters, defines the equations, and uses a numerical method to solve for the time evolution of the order parameter and magnetic field.

```python
import numpy as np
import matplotlib.pyplot as plt

def initialize_grid(N, L):
    '''
    Initialize the spatial grid.
    :param N: Number of grid points.
    :param L: Length of the domain.
    :return: Grid points and a zeroed order parameter.
    '''
    x = np.linspace(-L/2, L/2, N)
    return x, np.zeros(N, dtype=complex)

def update_fields(psi, A, alpha, beta, dt, dx):
    '''
    Update the order parameter and magnetic vector potential using
    ↪    the Ginzburg-Landau equations.
    :param psi: Current order parameter.
    :param A: Current vector potential.
    :param alpha: Material parameter.
    :param beta: Material parameter.
    :param dt: Time step for the simulation.
    :param dx: Spatial step size.
    :return: Updated order parameter and vector potential.
```

```python
    '''
    # Simple finite difference algorithm for updating psi and A
    nabla_psi = np.roll(psi, -1) - 2 * psi + np.roll(psi, 1)   #
    ↪  Second derivative in space
    J = -1j * (np.conj(psi) * (np.gradient(A)) + np.gradient(psi))
    ↪  # Current density calculation

    # Ginzburg-Landau equations
    psi_new = psi + dt * (alpha * psi + beta * np.abs(psi)**2 * psi
    ↪  + 0.5 * nabla_psi)
    A_new = A + dt * (1.0 / (4 * np.pi)) * (np.gradient(J))

    return psi_new, A_new

def simulate_ginzburg_landau(N, L, alpha, beta, dt, total_time):
    '''
    Simulate the Ginzburg-Landau dynamics for a superconducting
    ↪  system.
    :param N: Number of grid points.
    :param L: Length of the domain.
    :param alpha: Material parameter.
    :param beta: Material parameter.
    :param dt: Time step for the simulation.
    :param total_time: Total time to simulate.
    :return: Time evolution of the order parameter.
    '''
    x, psi = initialize_grid(N, L)
    A = np.zeros(N)
    num_steps = int(total_time / dt)
    history = []

    for step in range(num_steps):
        psi, A = update_fields(psi, A, alpha, beta, dt, (L/N))
        history.append(np.abs(psi))

    return x, history

# Parameters
N = 256            # Number of spatial grid points
L = 10.0           # Length of the domain
alpha = -1.0       # Material parameter
beta = 1.0         # Material parameter
dt = 0.01          # Time step
total_time = 5.0   # Total simulation time

# Run the simulation
x, history = simulate_ginzburg_landau(N, L, alpha, beta, dt,
↪  total_time)

# Plot the results
plt.figure(figsize=(8, 6))
for i in range(0, len(history), 10):   # Plot every 10th frame
    plt.plot(x, history[i], label=f'Time {i*dt:.2f}')
```

```
plt.title('Ginzburg-Landau Dynamics of Superconductivity')
plt.xlabel('Position')
plt.ylabel('Order Parameter |Ψ|')
plt.legend()
plt.grid()
plt.show()
```

This code implements the Ginzburg-Landau equations to model the dynamics of superconductivity in a given system.

- `initialize_grid` sets up the spatial grid and initializes the order parameter.
- `update_fields` computes the updated order parameter and magnetic vector potential based on the Ginzburg-Landau equations.
- `simulate_ginzburg_landau` runs the simulation and records the time evolution of the order parameter.

The simulation results are plotted, showing the evolution of the superconducting order parameter over time across the spatial domain. This approach provides a visual representation of how the order parameter evolves in response to the system dynamics.

Multiple Choice Questions

1. Who first proposed the Ginzburg-Landau theory of superconductivity?

    (a) Albert Einstein and Niels Bohr

    (b) Vitaly Ginzburg and Lev Landau

    (c) John Bardeen and Leon Cooper

    (d) Richard Feynman and Julian Schwinger

2. Which of the following best describes the order parameter Ψ in the Ginzburg-Landau equations?

    (a) It is a scalar function indicating energy loss.

    (b) It is a complex-valued function representing the density of superconducting pairs.

    (c) It is a real-valued function calculating temperature.

    (d) It is a vector function representing magnetic field strength.

3. What role does the vector potential **A** play in the Ginzburg-Landau equations?

(a) It defines material-specific properties.

(b) It describes the electric field in superconductors.

(c) It is related to the magnetic field and influences the phase of the order parameter.

(d) It calculates the temperature dependence in superconductive materials.

4. Vortex solutions in the context of the Ginzburg-Landau equations are associated with:

   (a) Zero electrical resistance.

   (b) Phase transitions in superconductors.

   (c) Quantized magnetic flux and topological defects in the order parameter.

   (d) None of the above.

5. When simulating the Ginzburg-Landau equations numerically, which of the following methods is commonly utilized?

   (a) Monte Carlo methods

   (b) Finite difference methods

   (c) Game theory analysis

   (d) Linear programming techniques

6. The Ginzburg-Landau theory primarily applies to which of the following areas?

   (a) Classical mechanics

   (b) Thermodynamics

   (c) Quantum mechanics, particularly superconductivity

   (d) Electromagnetism

7. What phenomenon can be investigated by analyzing the solutions to the Ginzburg-Landau equations?

   (a) Electron conductivity in metals

   (b) Critical temperature behavior of superconductors

   (c) Fluid dynamics in atmospheric science

   (d) Planetary motion in astrophysics

Answers:

1. **B: Vitaly Ginzburg and Lev Landau** Vitaly Ginzburg and Lev Landau formulated the theoretical framework of superconductivity known as the Ginzburg-Landau theory, which describes the behavior of superconductors through differential equations.

2. **B: It is a complex-valued function representing the density of superconducting pairs.** The order parameter Ψ captures the state of the superconducting phase and is crucial in determining the density of Cooper pairs within the material.

3. **C: It is related to the magnetic field and influences the phase of the order parameter.** The vector potential **A** is integral to representing magnetic effects in the Ginzburg-Landau formalism, affecting the dynamics of the superconductive state.

4. **C: Quantized magnetic flux and topological defects in the order parameter.** Vortex solutions represent regions where magnetic flux is quantized, indicating the presence of defects in the order parameter, which plays a significant role in vortex dynamics in superconductors.

5. **B: Finite difference methods** Due to the complexity and nonlinearity of the Ginzburg-Landau equations, numerical simulations often employ finite difference methods for approximating solutions.

6. **C: Quantum mechanics, particularly superconductivity** The Ginzburg-Landau theory is a pivotal aspect of condensed matter physics, specifically focused on quantum mechanical descriptions of superconductivity.

7. **B: Critical temperature behavior of superconductors** The solutions to the Ginzburg-Landau equations allow researchers to explore various superconductive characteristics, including the critical temperature at which materials transition into the superconducting state.

Chapter 19

Elliptic Integro-Differential Equations

In this chapter, we delve into the realm of elliptic integro-differential equations. These equations arise in a wide range of mathematical and physical problems, and their study involves a combination of differential and integral operators. We will explore the mathematical foundations of these equations, analytical techniques for their solutions, probabilistic interpretations, numerical methods for approximations, and various applications in financial mathematics.

Introduction and Basic Examples

Elliptic integro-differential equations combine differential operators with integral operators, making them more challenging to solve compared to purely differential equations. These equations often arise in problems involving the balance between diffusion and interaction effects. One of the fundamental examples is the Fredholm integral equation of the second kind, given by

$$u(x) - \int_a^b K(x,y)u(y)dy = f(x), \quad a \leq x \leq b,$$

where K is a given kernel function, f is a given function, and u is the unknown function to be determined.

Analytical Techniques

Analytical techniques for solving elliptic integro-differential equations heavily rely on the properties of integral operators and the theory of integral equations. A common approach is to use iterative methods, such as the Neumann series expansion, to approximate the solution of the integral equation. In addition, classical techniques such as separation of variables and Green's function methods can be employed to obtain analytical solutions for specific cases.

Probabilistic Interpretation

Elliptic integro-differential equations can also be interpreted probabilistically. These equations arise naturally in problems involving random walks or diffusions. In this context, the solution to the equation represents the expected value of a certain quantity at a given location or time. By formulating the equation in terms of probability densities and transition probabilities, one can connect the intuitive probabilistic interpretation to the mathematical formulation of the equations.

Numerical Methods

Numerical methods for solving elliptic integro-differential equations are essential when analytical solutions are not feasible. Finite difference methods, finite element methods, and spectral methods can be adapted to handle these equations. Additionally, there are specialized techniques, such as the Nyström method and the boundary element method, that exploit the integral nature of the equations and provide efficient numerical approximations.

Applications in Financial Mathematics

Elliptic integro-differential equations find significant applications in financial mathematics, particularly in option pricing and risk management. These equations arise in models such as the Black-Scholes equation and the Heston model, which are essential for pricing and hedging financial derivatives. By solving these equations, one can obtain option prices, implied volatilities, and other quantities of interest in the field of finance.

The study of elliptic integro-differential equations provides a deeper understanding of the interplay between differential and integral operators and their applications in various mathematical fields. By exploring analytical techniques, probabilistic interpretations, numerical methods, and applications in finance, researchers can unlock the potential of these equations in tackling complex problems. Further investigation of elliptic integro-differential equations will lead to advancements in mathematical theory and practical applications in finance and beyond.Sure! Below is a comprehensive Python code snippet that demonstrates key concepts related to elliptic integro-differential equations discussed in the chapter. We will create a function to solve the Fredholm integral equation of the second kind, implement numerical methods, and utilize those techniques for applications in financial mathematics.

"'latex

Python Code Snippet

Below is a Python code snippet that implements a basic solution to the Fredholm integral equation of the second kind using numerical methods.

```
import numpy as np
import matplotlib.pyplot as plt

def kernel_function(x, y):
    '''
    Example kernel function for the Fredholm integral equation.
    :param x: Variable x.
    :param y: Variable y.
    :return: Value of the kernel function.
    '''
    return np.exp(-np.abs(x - y))

def right_hand_side(x):
    '''
    Function representing the right-hand side of the Fredholm
    ↪   integral equation.
    :param x: Variable x.
    :return: Value of the right-hand side function f(x).
    '''
    return np.sin(np.pi * x)

def fredholm_integral_equation_solver(a, b, n):
    '''
    Solve the Fredholm integral equation of the second kind.
```

```
    :param a: Start of the interval.
    :param b: End of the interval.
    :param n: Number of discretization points.
    :return: Numerical solution of the equation.
    '''
    x = np.linspace(a, b, n)
    h = (b - a) / (n - 1)

    # Construct the system of equations
    K = np.zeros((n, n))
    for i in range(n):
        for j in range(n):
            K[i, j] = kernel_function(x[i], x[j]) * h

    f = np.array([right_hand_side(xi) * h for xi in x])

    u = np.linalg.solve(K, f)
    return x, u

# Parameters
a = 0    # Start of interval
b = 1    # End of interval
n = 100  # Number of discretization points

# Solve the Fredholm integral equation
x, solution = fredholm_integral_equation_solver(a, b, n)

# Plot the result
plt.plot(x, solution, label='Numerical Solution')
plt.title('Solution to the Fredholm Integral Equation of the Second
↪    Kind')
plt.xlabel('x')
plt.ylabel('u(x)')
plt.legend()
plt.grid()
plt.show()
```

This code defines several functions:

- kernel_function defines the kernel function used in the Fredholm integral equation.
- right_hand_side specifies the right-hand side function $f(x)$.
- fredholm_integral_equation_solver implements the numerical solution to the Fredholm integral equation by constructing and solving a linear system.

The implementation discretizes the interval $[a, b]$ into n points, constructs the integral operator matrix using the kernel function, and computes the solution. Finally, the solution is plotted to visualize the result.

This example illustrates how to tackle elliptic integro-differential equations using numerical methods, relevant in both mathematical theory and practical applications, particularly in finance.

Multiple Choice Questions

1. Which of the following describes an elliptic integro-differential equation?

    (a) An equation involving only differential operators

    (b) An equation that combines differential and integral operators

    (c) An equation involving only integral operators

    (d) An equation that is always linear

2. In what type of problems do elliptic integro-differential equations typically arise?

    (a) Problems involving purely time-dependent dynamics

    (b) Problems involving balance between diffusion and interaction

    (c) Problems defined on discrete domains only

    (d) Problems without boundary conditions

3. What method can be used for obtaining analytical solutions to elliptic integro-differential equations?

    (a) Separation of variables

    (b) Genetic algorithms

    (c) Game theory

    (d) Markov chains

4. The Neumann series expansion is best described as:

    (a) An iterative method to approximate solutions of integral equations

    (b) A method to find closed-form solutions

    (c) A numerical technique for solving linear ordinary differential equations

(d) A theoretical approach that does not apply to integro-differential equations

5. The probabilistic interpretation of elliptic integro-differential equations connects them to:

   (a) Transition probabilities in random walks

   (b) Nonlinear optimization problems

   (c) Deterministic systems only

   (d) Algebraic equations

6. Which numerical method is particularly suited for handling the integral nature of elliptic integro-differential equations?

   (a) Finite element method

   (b) Monte Carlo simulations

   (c) Finite difference method

   (d) Nyström method

7. In finance, elliptic integro-differential equations are commonly used in which context?

   (a) Portfolio optimization

   (b) Option pricing

   (c) Statistical inference

   (d) Risk-free interest rate models

Answers:

1. **B: An equation that combines differential and integral operators** Elliptic integro-differential equations are characterized by the combination of differential and integral operators, which makes them more complex than standard differential equations.

2. **B: Problems involving balance between diffusion and interaction** These equations often arise in contexts where there is a need to model interactions over a certain medium, typically where diffusion plays a significant role alongside interaction effects.

3. **A: Separation of variables** Separation of variables is a classical analytical technique for solving partial differential equations, which can also be adapted to find solutions to certain integro-differential equations.

4. **A: An iterative method to approximate solutions of integral equations** The Neumann series expansion is used as an iterative method to approximate solutions, particularly for integral equations, providing a series representation that can converge to an actual solution.

5. **A: Transition probabilities in random walks** The probabilistic interpretation links these equations to random processes, where the solutions represent expected values derived from the behavior of probabilistic models like random walks or diffusion processes.

6. **D: Nyström method** The Nyström method is specifically designed for integral equations and is particularly effective in obtaining approximate solutions for integro-differential equations, leveraging their integral components.

7. **B: Option pricing** In finance, elliptic integro-differential equations, such as those arising in the Black-Scholes framework, are utilized in the context of option pricing and derivatives management, making them crucial for theoretical and practical applications in finance.

Chapter 20

The Fokker-Planck Equation

The Fokker-Planck equation, also known as the Kolmogorov forward equation, is a partial differential equation that describes the evolution of the probability density function (PDF) of a stochastic process. It plays a crucial role in the field of statistical mechanics and has wide-ranging applications in various branches of science and engineering. In this chapter, we will explore the mathematical foundations of the Fokker-Planck equation and its time-harmonic case, as well as analytical and numerical methods for solving it. We will also discuss the long-time behavior and stationarity of solutions and delve into applications in statistical mechanics.

Stochastic Processes Background

Stochastic processes provide a probabilistic description of the time evolution of random phenomena. The Fokker-Planck equation specifically relates to continuous-time Markov processes, which are memoryless processes with continuous state spaces. Understanding the basics of stochastic processes is essential for comprehending the motivation and context behind the Fokker-Planck equation.

Derivation and Analytical Solutions

The Fokker-Planck equation can be derived using the Chapman-Kolmogorov equation and the diffusion approximation. It characterizes the time evolution of the PDF of a stochastic process by incorporating drift and diffusion terms. Analytical solutions to the Fokker-Planck equation are generally challenging to obtain due to its nonlinearity and complexity. However, for certain simplified cases, such as linear drift and diffusion coefficients, exact solutions can be found, providing insights into the behavior of the underlying stochastic process.

Numerical Methods

Numerical methods play a critical role in approximating solutions to the Fokker-Planck equation when analytical solutions are not feasible. Finite difference methods, finite element methods, and spectral methods can be adapted to discretize the equation and solve it numerically. The choice of numerical method depends on the specific characteristics of the problem at hand, such as the smoothness of the coefficients and the desired accuracy.

Long-Time Behavior and Stationarity

The study of the long-time behavior and stationary solutions of the Fokker-Planck equation is of fundamental importance in understanding the asymptotic dynamics and equilibrium states of stochastic processes. Stationary solutions correspond to time-independent PDFs and often provide insights into the statistical properties of the system under consideration. Analytical and numerical techniques can be employed to determine the existence and stability of such stationary solutions.

Applications in Statistical Mechanics

The Fokker-Planck equation finds widespread applications in the field of statistical mechanics. It serves as a powerful tool for studying the dynamics of particles in physical and chemical systems, diffusion processes, and the behavior of complex systems. Various phenomena, such as Brownian motion, random walks, and relaxation processes, can be effectively modeled and analyzed using the Fokker-Planck equation. The equation's applications extend to fields such as molecular dynamics, population dynamics, and the study of collective behavior in biological systems.

The Fokker-Planck equation provides a mathematical framework for analyzing the time evolution and statistical properties of stochastic processes. Its derivation, analytical solutions, numerical methods, and applications in statistical mechanics offer valuable insights into a wide range of phenomena. By studying the Fokker-Planck equation, researchers deepen their understanding of the behavior of random processes and gain the necessary tools for tackling complex problems in diverse scientific and engineering domains.

Python Code Snippet

Below is a Python code snippet that implements the Fokker-Planck equation using numerical methods to approximate the probability density function (PDF) of a stochastic process. This code demonstrates the finite difference method for solving the time evolution of the PDF.

```python
import numpy as np
import matplotlib.pyplot as plt

def initialize_pdf(x, initial_condition):
    '''
    Initialize the probability density function (PDF) based on the
      given initial condition.
    :param x: Array of state values.
    :param initial_condition: Function defining the initial
      condition.
    :return: Array representing the initial PDF.
    '''
    return initial_condition(x)

def finite_difference_fokker_planck(x, t_max, dt, D, mu,
    initial_condition):
    '''
    Solve the Fokker-Planck equation using the finite difference
      method.
    :param x: Array of state values.
    :param t_max: Maximum time for the simulation.
    :param dt: Time step.
    :param D: Diffusion coefficient.
    :param mu: Drift coefficient.
    :param initial_condition: Function defining the initial
      condition.
```

```python
    :return: Array containing the PDF at each time step.
    '''
    n = len(x)
    n_timesteps = int(t_max / dt)
    pdf = np.zeros((n_timesteps, n))

    # Initialize the PDF
    pdf[0, :] = initialize_pdf(x, initial_condition)

    # Time-stepping loop
    for t in range(1, n_timesteps):
        for i in range(1, n - 1):
            pdf[t, i] = pdf[t-1, i] + dt * (D * (pdf[t-1, i+1] - 2 *
            ↪   pdf[t-1, i] + pdf[t-1, i-1]) / (x[1] - x[0])**2
                        - mu * (pdf[t-1, i] - pdf[t-1, i-1]) / (x[1] -
            ↪   x[0]))

    return pdf

# Parameters
x_min = -5
x_max = 5
n_points = 100
x = np.linspace(x_min, x_max, n_points)

t_max = 1.0   # Maximum time
dt = 0.01     # Time step
D = 0.1       # Diffusion coefficient
mu = 0.05     # Drift coefficient

# Define the initial condition
def initial_condition(x):
    return np.exp(-x**2) / np.sqrt(np.pi)   # Gaussian initial
    ↪   condition

# Solve the Fokker-Planck equation
pdf_solution = finite_difference_fokker_planck(x, t_max, dt, D, mu,
↪   initial_condition)

# Plotting the results
for t in range(0, int(t_max / dt), 10):   # Plot every 10 time steps
    plt.plot(x, pdf_solution[t, :], label=f'Time = {t*dt:.2f}')

plt.title("Fokker-Planck Equation Solution")
plt.xlabel("State Space")
plt.ylabel("Probability Density Function (PDF)")
plt.legend()
plt.grid()
plt.show()
```

This code defines the following functions:

- `initialize_pdf` initializes the PDF using a specified initial condition.
- `finite_difference_fokker_planck` numerically solves the Fokker-Planck equation using a finite difference approach.

The main portion of the code sets parameters for the state space, maximum simulation time, diffusion and drift coefficients, and the initial condition, computes the PDF over time, and finally plots the results showing the evolution of the probability density function as a function of state space and time.

Multiple Choice Questions

1. The Fokker-Planck equation primarily describes the evolution of which of the following?

    (a) The average value of a random process

    (b) The time-dependent probability density function

    (c) The variance of a stochastic process

    (d) The autocorrelation function of a random variable

2. Which of the following terms is NOT typically associated with the Fokker-Planck equation?

    (a) Drift term

    (b) Diffusion term

    (c) Viscosity term

    (d) Probability density function

3. A stationary solution of the Fokker-Planck equation is characterized by which of the following properties?

    (a) It evolves over time towards a constant value.

    (b) It remains unchanged over time.

    (c) It oscillates between different values over time.

    (d) It does not exist for most systems.

4. Which numerical method is commonly used for solving the Fokker-Planck equation?

    (a) Euler's method

    (b) Monte Carlo simulations

(c) Finite difference methods

(d) Laplace transform method

5. The Fokker-Planck equation can be derived from which of the following principles?

    (a) Fourier analysis

    (b) The Chapman-Kolmogorov equation

    (c) The Law of Large Numbers

    (d) The Central Limit Theorem

6. In which field does the Fokker-Planck equation find significant applications?

    (a) Topology

    (b) Statistical mechanics

    (c) Number theory

    (d) Game theory

7. The Fokker-Planck equation is classified as:

    (a) A linear ordinary differential equation

    (b) A nonlinear hyperbolic partial differential equation

    (c) A linear partial differential equation

    (d) A nonlinear elliptic partial differential equation

Answers:
1. **B: The time-dependent probability density function** The Fokker-Planck equation specifically models the dynamics of the probability density function of a stochastic process over time, capturing how probabilities evolve.

2. **C: Viscosity term** The viscosity term is not a component of the Fokker-Planck equation. Instead, the equation includes a drift term representing systematic movements and a diffusion term representing random movements.

3. **B: It remains unchanged over time.** A stationary solution to the Fokker-Planck equation implies that the probability density function does not change as time evolves, indicating a balance in the rates of inflow and outflow of probability.

4. **C: Finite difference methods** Finite difference methods are one of the standard numerical techniques used for discretizing

and solving the Fokker-Planck equation when analytical solutions are not possible.

5. **B: The Chapman-Kolmogorov equation** The Fokker-Planck equation's derivation involves the Chapman-Kolmogorov equation which details how probabilities evolve over time in stochastic processes.

6. **B: Statistical mechanics** The Fokker-Planck equation has extensive applications in statistical mechanics, as it helps in understanding the behavior of particles and systems at equilibrium.

7. **C: A linear partial differential equation** The Fokker-Planck equation is classified as a linear partial differential equation because it can typically be expressed in a linear form with respect to the probability density function.

Chapter 21

The Fisher-KPP Equation

In this chapter, we delve into the intricacies of the Fisher-KPP equation, named after Ronald Fisher, Alexander Kolmogorov, and Andrey Petrovich Piskunov. This equation is a fundamental model in population genetics and mathematical biology, used to analyze the spread of advantageous traits or genes in a population. We will explore the phase transition phenomenon described by this equation, investigate its traveling wave solutions, discuss various analytical techniques for studying them, and showcase numerical simulations to gain insights into the dynamics of this equation.

Population Genetics Context

The Fisher-KPP equation emerges as a model in population genetics, aiming to understand the distribution and spread of advantageous traits within a population. It incorporates the effects of both genetic drift and natural selection. By analyzing this equation, we can gain valuable insights into the evolutionary dynamics of populations.

Traveling Wave Solutions

The Fisher-KPP equation exhibits a fascinating phenomenon known as phase transition, wherein a localized perturbation spreads throughout an entire population. This behavior is captured by traveling wave solutions of the equation, which describe the propagation of the advantageous trait over time. These solutions capture the dynamics of the population as well as its speed of invasion.

Analytical Techniques

Analyzing the Fisher-KPP equation often involves challenging mathematical techniques. We explore various methods employed by researchers to study the equation and derive analytical results. These techniques include linear stability analysis, asymptotic analysis, and transformation methods. Through these methodologies, we can unravel the properties of the traveling wave solutions and gain deeper insights into the underlying dynamics.

Numerical Simulations

Numerical simulations play a crucial role in understanding the Fisher-KPP equation, as analytical solutions are often limited to special cases. We discuss numerical methods and algorithms used to approximate the traveling wave solutions. Finite difference methods, finite element methods, and spectral methods are commonly employed to numerically solve the Fisher-KPP equation. These simulations allow us to visualize the behavior of the traveling waves and explore parameter dependencies.

Applications in Epidemiology

The Fisher-KPP equation finds significant applications in epidemiology, specifically in modeling the spread of infectious diseases. By considering the population dynamics and the transmission rate of the disease, the Fisher-KPP equation can provide insights into the epidemic's progression. Analyzing the traveling wave solutions can help in understanding the speed and extent of the disease's invasion, aiding in the implementation of appropriate control measures.

The Fisher-KPP equation presents a powerful mathematical tool for studying population genetics and analyzing the spread of advantageous traits or genes within a population. Through the exploration of its traveling wave solutions, application of analytical techniques, and utilization of numerical simulations, we can gain a deeper understanding of the dynamics and implications of this equation. This knowledge is invaluable not only in the field of population genetics but also extends to broader applications in epidemiology and other areas of mathematical biology.Here is a comprehensive Python code snippet that implements key elements associated with the Fisher-KPP equation, including the derivation of traveling wave solutions and numerical simulation using finite difference methods. This code will help demonstrate how to work with the Fisher-KPP equation in the context of population dynamics.

Python Code Snippet

Below is a Python code snippet that models the Fisher-KPP equation using a finite difference method to approximate the traveling wave solutions.

```
import numpy as np
import matplotlib.pyplot as plt

def initialize_population_grid(size, initial_population_fraction):
    '''
    Initialize the population grid with a given size and initial
    ↪ population fraction.
    :param size: Size of the grid (number of cells).
    :param initial_population_fraction: Fraction of the grid cells
    ↪ initially populated.
    :return: Initial population distribution.
    '''
    grid = np.zeros(size)
    # Set initial condition: localized population in the center
    mid = size // 2
    grid[mid - int(size * initial_population_fraction / 2):mid +
    ↪ int(size * initial_population_fraction / 2)] = 1
    return grid

def update_population(grid, diffusion_rate, reproduction_rate, dt):
    '''
    Update the population grid using the Fisher-KPP equation.
    :param grid: Current population distribution.
    :param diffusion_rate: Rate of diffusion of the population.
    :param reproduction_rate: Rate of reproduction of the
    ↪ population.
    :param dt: Time step for the simulation.
    :return: Updated population distribution.
    '''
    new_grid = grid.copy()
```

```python
    size = len(grid)

    for i in range(1, size - 1):
        # Fisher-KPP equation finite difference approximation
        new_grid[i] = grid[i] + dt * (diffusion_rate * (grid[i + 1]
            - 2 * grid[i] + grid[i - 1]) + reproduction_rate *
            grid[i] * (1 - grid[i]))

    # Neumann boundary condition (zero-flux)
    new_grid[0] = new_grid[1]
    new_grid[-1] = new_grid[-2]

    return new_grid

def simulate_fisher_kpp(iterations, size, diffusion_rate,
    reproduction_rate, dt, initial_population_fraction):
    '''
    Simulate the Fisher-KPP equation over a number of iterations.
    :param iterations: Number of time steps to simulate.
    :param size: Size of the population grid.
    :param diffusion_rate: Rate of population diffusion.
    :param reproduction_rate: Rate of population reproduction.
    :param dt: Time step for the simulation.
    :param initial_population_fraction: Initial population fraction
        in the grid.
    :return: Array of population distributions over time.
    '''
    population_distribution = np.zeros((iterations, size))
    population_distribution[0] = initialize_population_grid(size,
        initial_population_fraction)

    for t in range(1, iterations):
        population_distribution[t] =
            update_population(population_distribution[t - 1],
            diffusion_rate, reproduction_rate, dt)

    return population_distribution

# Parameters
iterations = 200  # Total time steps
size = 100  # Size of the grid
diffusion_rate = 0.1  # Population diffusion rate
reproduction_rate = 1.0  # Population reproduction rate
dt = 0.01  # Time step size
initial_population_fraction = 0.1  # Initial fraction of population

# Run simulation
population_distribution = simulate_fisher_kpp(iterations, size,
    diffusion_rate, reproduction_rate, dt,
    initial_population_fraction)

# Plotting results
plt.figure(figsize=(10, 6))
```

```
for t in range(0, iterations, 20):  # Plot every 20 time steps
    plt.plot(population_distribution[t], label=f'Time step {t}')
plt.title('Fisher-KPP Equation Simulation')
plt.xlabel('Grid position')
plt.ylabel('Population density')
plt.legend()
plt.grid()
plt.show()
```

This code defines several functions:

- `initialize_population_grid` initializes a 1D grid with a specified fraction of the population.
- `update_population` performs a finite difference update based on the Fisher-KPP equation.
- `simulate_fisher_kpp` runs the simulation for a defined number of iterations and returns the population distribution over time.

The simulated population dynamics are visualized through a plot depicting population density across the grid at specific time steps, demonstrating how advantageous traits spread in a population over time.

Multiple Choice Questions

1. What is the primary focus of the Fisher-KPP equation?

    (a) Modeling fluid dynamics

    (b) Analyzing population genetics and the spread of advantageous traits

    (c) Solving linear algebraic equations

    (d) Predicting stock market behavior

2. The Fisher-KPP equation is named after which of the following individuals?

    (a) Isaac Newton, Albert Einstein, Richard Feynman

    (b) Ronald Fisher, Alexander Kolmogorov, Andrey Piskunov

    (c) Charles Darwin, Gregor Mendel, Stephen Jay Gould

    (d) Claude Shannon, Alan Turing, John von Neumann

3. What mathematical phenomenon is associated with traveling wave solutions in the context of the Fisher-KPP equation?

(a) Interference patterns

(b) Phase transition

(c) Nonlinear oscillations

(d) Chaotic behavior

4. Which numerical method is commonly utilized to approximate solutions of the Fisher-KPP equation?

    (a) Runge-Kutta method

    (b) Finite difference method

    (c) Gradient descent method

    (d) Newton's method

5. In epidemiology, the Fisher-KPP equation can help model what aspect of infectious diseases?

    (a) The financial costs of outbreaks

    (b) The genetic diversity of pathogens

    (c) The spread and invasion dynamics of diseases

    (d) The vaccine development process

6. Which of the following techniques is NOT typically used in the analysis of the Fisher-KPP equation?

    (a) Linear stability analysis

    (b) Asymptotic analysis

    (c) Bootstrap methods

    (d) Transformation methods

7. True or False: The Fisher-KPP equation is primarily applicable to static systems without feedback.

    (a) True

    (b) False

Answers:

1. **B: Analyzing population genetics and the spread of advantageous traits** The Fisher-KPP equation serves primarily as a model for understanding how advantageous traits spread within populations in the field of population genetics.

2. **B: Ronald Fisher, Alexander Kolmogorov, Andrey Piskunov** The equation is named after these three prominent figures who contributed to the formulation of the model used in population genetics.

3. **B: Phase transition** Traveling wave solutions in the Fisher-KPP equation demonstrate the phenomenon of phase transition, where a localized trait can invade the population over time.

4. **B: Finite difference method** The finite difference method is a commonly employed numerical technique used to approximate solutions to the Fisher-KPP equation, among other PDEs.

5. **C: The spread and invasion dynamics of diseases** In epidemiology, the Fisher-KPP equation provides insights into how infectious diseases spread, allowing for the exploration of various intervention strategies.

6. **C: Bootstrap methods** Bootstrap methods are generally used for statistical inference, rather than analyzing the dynamics of the Fisher-KPP equation, which employs methods like stability and asymptotic analyses.

7. **B: False** The Fisher-KPP equation deals with dynamic systems where feedback and population interactions are central to understanding the evolution of traits over time, contrary to the claim of it being applicable to static systems.

Chapter 22

The Fractional Laplacian

The study of partial differential equations (PDEs) has been crucial in understanding various physical phenomena. The Laplace operator, defined as the divergence of the gradient of a function, is a fundamental operator in PDEs, describing behaviors such as diffusion and heat conduction. In certain cases, however, the conventional Laplacian may not fully capture the underlying dynamics of the system. The fractional Laplacian provides a powerful extension of the Laplace operator by introducing nonlocality, enabling the modeling of anomalous diffusion and other phenomena where long-range interactions play a significant role. In this chapter, we explore the foundations and applications of the fractional Laplacian, showcasing its importance in the realm of anomalous diffusion.

Definition and Properties

The fractional Laplacian, denoted as Δ^s for $s \in (0, 1)$, is a nonlocal operator that generalizes the conventional Laplacian to incorporate fractional-order derivatives. It is defined through the Fourier transform as follows:

$$\mathcal{F}\left\{\Delta^s u\right\}(\xi) = -|\xi|^{2s} \mathcal{F}\{u\}(\xi), \tag{22.1}$$

where $\mathcal{F}\{u\}$ represents the Fourier transform of the function u, and $\xi \in \mathbb{R}^d$ is the frequency variable. The fractional Laplacian possesses several key properties that make it an indispensable tool in various mathematical and physical contexts, including fractional-order differentiation, nonlocality, and self-adjointness.

Spectral Representation

The spectral representation of the fractional Laplacian provides insights into its eigenvalues and eigenfunctions. By utilizing the Mittag-Leffler function, which generalizes the exponential function, we can express the eigenvalues and eigenfunctions of the fractional Laplacian in integral form. The spectral decomposition allows us to study the behavior of the fractional Laplacian and its applications in the analysis of PDEs involving the fractional Laplacian.

Numerical Approximations

Computing solutions involving the fractional Laplacian often requires numerical approximations due to the nonlocal and integral nature of the operator. Various numerical methods have been developed to approximate the fractional Laplacian, including the classical Grünwald-Letnikov scheme, the Fourier spectral method, and the finite difference method. These approaches provide options for approximating the fractional Laplacian and obtaining numerically accurate solutions for PDEs involving this operator.

Boundary Value Problems

Solving boundary value problems involving the fractional Laplacian requires careful consideration of the appropriate boundary conditions. Analogous to the classical Laplacian, the fractional Laplacian admits different types of boundary conditions, such as Dirichlet, Neumann, and Robin boundary conditions. Understanding the implications and properties of these boundary conditions is essential for the well-posedness and stability of solutions to PDEs involving the fractional Laplacian.

Applications in Anomalous Diffusion

Anomalous diffusion phenomena are prevalent in various fields, including physics, biology, and finance. The fractional Laplacian provides a powerful mathematical tool for modeling and analyzing systems exhibiting anomalous diffusion. By incorporating the nonlocal and long-range nature of interactions, the fractional Laplacian enables the accurate description of various anomalous diffusion processes, such as Levy flights, subdiffusion, and superdiffusion. The understanding and application of the fractional Laplacian in anomalous diffusion scenarios have far-reaching implications in fields including material science, transport phenomena, and stochastic processes.

The fractional Laplacian serves as a versatile operator, demonstrating its utility and significance beyond the classical Laplacian. By incorporating nonlocality and fractional-order derivatives, it provides a powerful framework for modeling anomalous diffusion and elucidating the underlying dynamics of diverse phenomena. The properties, spectral representation, numerical approximation techniques, boundary value problems, and applications of the fractional Laplacian explored in this chapter offer invaluable insights into the behavior of complex systems in fields ranging from physics to biology and beyond.

Python Code Snippet

Below is a Python code snippet that implements key equations and algorithms related to the fractional Laplacian, including its definition, spectral representation, and numerical approximation methods.

```
import numpy as np
import scipy.fftpack
import matplotlib.pyplot as plt

def fractional_laplacian(u, s, L):
    '''
    Compute the fractional Laplacian of a 1D function using Fourier
    ↪ Transforms.
    :param u: Input function as a NumPy array.
    :param s: Order of the fractional Laplacian (0 < s < 1).
    :param L: Length of the domain.
    :return: NumPy array representing the fractional Laplacian of
    ↪ the input function.
```

```python
    '''
    N = len(u)
    k = np.fft.fftfreq(N, L/N) * 2 * np.pi  # Frequency domain
    u_hat = np.fft.fft(u)  # FFT of the input function
    return np.fft.ifft(-np.abs(k)**(2*s) * u_hat).real  # Inverse
    ↪     FFT of the fractional Laplacian

def plot_fractional_laplacian():
    '''
    Plot the effect of the fractional Laplacian on an initial
    ↪    condition.
    '''
    L = 10  # Domain length
    x = np.linspace(0, L, 256)  # Spatial domain
    u = np.sin(np.pi * x / L)  # Initial function

    # Compute fractional Laplacian with different orders
    s_values = [0.25, 0.5, 0.75]
    plt.figure(figsize=(12, 8))
    for s in s_values:
        u_lap_frac = fractional_laplacian(u, s, L)
        plt.plot(x, u_lap_frac, label=f'Fractional Laplacian
        ↪    (s={s})')

    plt.title('Fractional Laplacian Effect')
    plt.xlabel('x')
    plt.ylabel('u')
    plt.legend()
    plt.grid()
    plt.show()

def numerical_solution_fractional_diffusion():
    '''
    Numerical solution of the fractional diffusion equation using
    ↪    finite differences.
    '''
    import numba  # For optimized performance on the loop

    def fractional_heat_equation(u, s, dt, dx, steps):
        for _ in range(steps):
            u_new = np.zeros_like(u)
            for i in range(1, len(u)-1):
                u_new[i] = u[i] + dt * fractional_laplacian(u, s,
                ↪    dx)
            u = u_new
        return u

    # Parameters
    L = 10.0
    dx = 0.1
    dt = 0.01
    steps = 100
    x = np.arange(0, L+dx, dx)
```

```
u = np.exp(-((x - 5) ** 2) / 1)    # Initial condition

# Solve fractional diffusion equation
u_final = fractional_heat_equation(u, s=0.5, dt=dt, dx=dx,
↪   steps=steps)   # s=0.5 for example

# Plot results
plt.figure(figsize=(8, 5))
plt.plot(x, u_final, label='Final State after Fractional
↪   Diffusion')
plt.title('Fractional Diffusion Equation Solution')
plt.xlabel('x')
plt.ylabel('u')
plt.legend()
plt.grid()
plt.show()

# Execute plotting functions
plot_fractional_laplacian()
numerical_solution_fractional_diffusion()
```

This code defines three functions:

- `fractional_laplacian` calculates the fractional Laplacian of a given function using the Fourier transform.
- `plot_fractional_laplacian` generates plots to visualize the effect of the fractional Laplacian on an initial function.
- `numerical_solution_fractional_diffusion` implements a numerical approach to solve the fractional diffusion equation using finite differences.

The provided code demonstrates the computation of the fractional Laplacian, visualizes its effects, and solves the fractional diffusion equation numerically, showcasing the operator's potential applications in modeling complex diffusion processes.

Multiple Choice Questions

1. What is the primary definition of the fractional Laplacian Δ^s?

    (a) A local derivative operator acting on smooth functions

    (b) A nonlocal operator defined through the Fourier transform

    (c) A convolution operator involving delta functions

(d) An extension of the gradient operator to non-integer dimensions

2. Which of the following properties is NOT associated with the fractional Laplacian?

    (a) Self-adjointness
    (b) Nonlocality
    (c) Linearity
    (d) Symmetry with respect to the mixed derivatives

3. In which context is the fractional Laplacian primarily used?

    (a) Traditional heat diffusion processes
    (b) Anomalous diffusion and transport phenomena
    (c) Classical fluid dynamics
    (d) Linear elasticity theory

4. What numerical method is commonly employed to approximate the fractional Laplacian?

    (a) Finite element method
    (b) Crank-Nicolson scheme
    (c) Grünwald-Letnikov scheme
    (d) Laplace smoothing

5. Which boundary condition can be applied to problems involving the fractional Laplacian?

    (a) Dirichlet boundary conditions only
    (b) Neumann boundary conditions only
    (c) Robin boundary conditions only
    (d) All types of boundary conditions (Dirichlet, Neumann, Robin)

6. The nonlocal behavior of the fractional Laplacian signifies:

    (a) Its derivation depends solely on local information
    (b) Its outcomes depend on an integral of values over a range of distances
    (c) Its invariance under translations

(d) It is applicable only in one-dimensional settings

7. One of the applications of the fractional Laplacian in real-world scenarios is:

   (a) Fluid flow in laminar regimes

   (b) Modeling subdiffusion processes in anomalous dynamics

   (c) Predicting outcome in classical mechanics

   (d) Heat dissipation in conductive materials

Answers:

1. **B: A nonlocal operator defined through the Fourier transform** The fractional Laplacian is a nonlocal operator that generalizes the classical Laplacian by incorporating fractional-order derivatives, and it is defined in terms of the Fourier transform.

2. **D: Symmetry with respect to the mixed derivatives** While the fractional Laplacian is self-adjoint, linear, and nonlocal, the property regarding symmetry in mixed derivatives is not a standard characterization of the fractional Laplacian.

3. **B: Anomalous diffusion and transport phenomena** The fractional Laplacian is particularly useful in modeling anomalous diffusion processes, where conventional models fail to accurately represent the dynamics of the system.

4. **C: Grünwald-Letnikov scheme** The Grünwald-Letnikov scheme is a classical numerical method used to approximate the fractional Laplacian, which leverages a finite difference approach to handle nonlocal operators.

5. **D: All types of boundary conditions (Dirichlet, Neumann, Robin)** The fractional Laplacian can be applied under various boundary conditions, which are required to establish well-posed problems within its framework.

6. **B: Its outcomes depend on an integral of values over a range of distances** The nonlocal nature of the fractional Laplacian means that the effect at a point is influenced by values of the function over a wider region, not just at the point itself.

7. **B: Modeling subdiffusion processes in anomalous dynamics** One significant application of the fractional Laplacian is in modeling subdiffusion, a process where the diffusion rate is slower than what is described by classical models, highlighting its utility in various scientific contexts.

Chapter 23

Elliptic Systems of Equations

In this chapter, we delve into the theory and analysis of elliptic systems of equations. Elliptic systems play a crucial role in various mathematical disciplines and physical phenomena, offering a rich field for exploration and study. The chapter focuses on the fundamental aspects of elliptic systems, including their mathematical setup, operator theory, Sobolev spaces for systems, numerical methods, and applications in multiphysics problems.

Coupled PDE Systems

Elliptic systems of equations consist of a set of partial differential equations that are coupled together. These systems commonly arise in mathematical modeling scenarios involving multiple interacting physical quantities or phenomena. Typical examples include the Navier-Stokes equations for fluid flow, the Maxwell's equations for electromagnetism, and the coupled systems in elasticity theory. Such systems exhibit intricate mathematical structures and demand advanced analytical and numerical techniques to discern meaningful solutions.

Operator Theory

The analysis of elliptic systems heavily relies on operator theory, which provides a rigorous mathematical framework for studying linear and nonlinear operators between function spaces. The understanding of operator theory aids in establishing well-posedness, existence, and uniqueness results for solutions of elliptic systems. Key concepts in operator theory include the Fredholm operators, compact operators, and the spectrum of operators, which enable the investigation of important qualitative properties of elliptic systems.

Sobolev Spaces for Systems

Sobolev spaces serve as the natural function spaces for studying elliptic systems. These spaces comprise functions and their derivatives that possess certain integrability conditions. The properties of Sobolev spaces play a vital role in establishing existence and regularity results for solutions of elliptic systems. The extension of Sobolev spaces to systems requires careful consideration of the interplay between different variables and equations, often involving additional regularity requirements on the solutions.

Numerical Methods

Numerical methods provide indispensable tools for approximating solutions of elliptic systems, owing to the complexity and nonlinearity involved. Finite element methods, finite difference methods, and spectral methods are commonly employed in the numerical treatment of elliptic systems. These approaches discretize the domain, approximate the unknown functions, and transform the original system into a system of algebraic equations, facilitating efficient computations and convergence to accurate solutions.

Applications in Multiphysics Problems

Elliptic systems find widespread application in multiphysics problems, where multiple physical phenomena interact and influence each other. The Navier-Stokes equations, governing fluid dynamics, represent one such example where the coupling occurs between the fluid velocity and pressure fields. Other examples include fluid-structure interaction problems, electromechanical systems, and multiphase flow problems. Understanding and solving elliptic systems are critical in elucidating the behavior of these complex and interconnected physical systems.

The study of elliptic systems of equations provides profound insights into the mathematical and physical principles underlying a variety of phenomena. The analysis of coupled PDE systems, employing operator theory and Sobolev spaces, enables the study of well-posedness and qualitative properties of solutions. Numerical methods become indispensable in practical scenarios, while applications in multiphysics highlight the significance of elliptic systems in addressing real-world challenges. This chapter serves as a comprehensive guide to the fundamental aspects of elliptic systems, offering a solid foundation for further exploration in this captivating field of study.Certainly! Below is the comprehensive Python code snippet that implements key concepts related to elliptic systems of equations, including functions to solve a coupled elliptic system using the finite element method. This section is presented using the minted package for LaTeX formatting.

Python Code Snippet

Below is a Python code snippet that demonstrates the finite element method for solving a simple elliptic system of equations represented by:

$$-\Delta u = f_1 \text{ in } \Omega$$
$$-\Delta v = f_2 \text{ in } \Omega$$
$$u = g \text{ on } \partial\Omega$$

The following Python code uses the FEniCS library to discretize and solve this elliptic system.

```
from dolfin import *

def solve_elliptic_system(f1, f2, g, mesh_size=32):
    '''
    Solve a coupled elliptic system in the form of -u = f1, -v = f2
    :param f1: Source term for the first equation.
    :param f2: Source term for the second equation.
    :param g: Boundary condition for the first equation.
    :param mesh_size: Number of divisions for the mesh.
```

```python
    :return: The solutions u and v.
    '''
    # Create mesh and define function space
    mesh = UnitSquareMesh(mesh_size, mesh_size)
    V = FunctionSpace(mesh, 'P', 1)  # Polynomial degree 1

    # Define boundary condition
    bc = DirichletBC(V, g, 'on_boundary')

    # Define test and trial functions
    u = TrialFunction(V)
    v = TrialFunction(V)
    w = TestFunction(V)

    # Define variational problem for u
    a1 = inner(grad(u), grad(w))*dx
    L1 = f1 * w * dx

    # Define variational problem for v
    a2 = inner(grad(v), grad(w))*dx
    L2 = f2 * w * dx

    # Assemble matrices and solve
    A1 = assemble(a1)
    A2 = assemble(a2)

    b1 = assemble(L1)
    b2 = assemble(L2)

    # Apply boundary conditions
    bc.apply(A1, b1)
    bc.apply(A2, b2)

    # Solve problems
    u_sol = Function(V)
    v_sol = Function(V)

    solve(A1, u_sol.vector(), b1)
    solve(A2, v_sol.vector(), b2)

    return u_sol, v_sol

# Example source terms and boundary condition
f1 = Expression('sin(pi*x[0])*sin(pi*x[1])', degree=2)
f2 = Expression('cos(pi*x[0])*cos(pi*x[1])', degree=2)
g = Constant(0)  # Dirichlet boundary condition

# Solve the elliptic system
u_solution, v_solution = solve_elliptic_system(f1, f2, g)

# Plotting the solutions
import matplotlib.pyplot as plt
plot(u_solution)
```

```
plt.title("Solution u")
plt.show()

plot(v_solution)
plt.title("Solution v")
plt.show()
```

This code defines the function `solve_elliptic_system`, which:

- Takes two source terms and a boundary condition.
- Creates a mesh and finite element function space.
- Sets up the variational forms for both equations and assembles the corresponding matrices.
- Solves the resulting linear systems for u and v.
- Finally, it visualizes the solutions using Matplotlib.

This implementation serves as a practical introduction to solving elliptic systems using the finite element method, providing a valuable computational tool for applied mathematics and engineering problems involving such equations.

Multiple Choice Questions

1. What is a characteristic feature of elliptic systems of equations?

    (a) They typically describe hyperbolic phenomena.

    (b) They consist of a single equation in one variable.

    (c) They are coupled sets of partial differential equations.

    (d) They are always time-dependent.

2. Which operator is widely used in the analysis of elliptic systems?

    (a) The Laplace operator

    (b) The Fourier operator

    (c) The Hessian operator

    (d) The divergence operator

3. Sobolev spaces are essential for:

    (a) Ensuring solutions are Lipschitz continuous.

    (b) Establishing existence and regularity of solutions.

(c) Providing only numerical solutions to PDEs.

(d) Eliminating boundary conditions in PDEs.

4. Which numerical method is most frequently used to solve elliptic systems?

   (a) Monte Carlo methods

   (b) Finite element methods

   (c) Particle-in-cell methods

   (d) Markov chain methods

5. In multiphysics problems, elliptic systems are commonly used to study:

   (a) Nonlinear dynamics without coupling effects.

   (b) The interaction of multiple physical phenomena.

   (c) Only thermal conduction in solids.

   (d) The diffusion of particles in isolation.

6. Which of the following is true regarding the well-posedness of elliptic systems?

   (a) It only pertains to linear equations.

   (b) It ensures that solutions exist, are unique, and depend continuously on data.

   (c) It is unnecessary for practical applications.

   (d) It only applies to systems without boundary conditions.

7. The study of operator theory in elliptic systems is primarily concerned with:

   (a) The geometric interpretation of PDEs.

   (b) The relationships between input and output functions.

   (c) Finding numerical solutions to linear equations.

   (d) The classification of oscillatory behavior.

Answers:
1. **C: They are coupled sets of partial differential equations.** Elliptic systems consist of multiple PDEs that are coupled together, which distinguishes them from simpler forms of PDEs that might only involve a single equation.

2. **A: The Laplace operator** The Laplace operator is a fundamental operator in the analysis of elliptic equations, widely used for its role in defining boundary value problems associated with these systems.

3. **B: Establishing existence and regularity of solutions.** Sobolev spaces provide a framework to examine the properties of solutions of PDEs, enabling the establishment of their existence and regularity under appropriate conditions.

4. **B: Finite element methods** Finite element methods are extensively used to numerically solve elliptic systems, offering flexibility and accuracy in handling complex geometries and boundary conditions.

5. **B: The interaction of multiple physical phenomena.** Elliptic systems are ideal for modeling multiphysics problems, where various physical processes interact, such as fluid-structure interactions and electromagnetism.

6. **B: It ensures that solutions exist, are unique, and depend continuously on data.** Well-posedness is characterized by existence, uniqueness, and continuous dependence on initial and boundary conditions, which is crucial for the stability of solutions.

7. **B: The relationships between input and output functions.** Operator theory focuses on understanding operators that act on function spaces, which is essential in analyzing elliptic systems for studying linear and nonlinear interactions.

Chapter 24

The Lichnerowicz Equation

Relativity Theory Context

In the field of relativity theory, the Lichnerowicz equation plays a fundamental role in the study of the geometry of spacetime and the behavior of gravitational fields. Named after the French mathematician André Lichnerowicz, this equation arises in the context of the Einstein field equations, which describe the interactions between matter, energy, and curvature in four-dimensional spacetime.

Conformal Metrics

The Lichnerowicz equation is concerned with the analysis of conformal metrics, which are metrics that differ from a given metric by a conformal factor. More precisely, for a given Riemannian metric g, the Lichnerowicz equation relates the conformal factor ϕ to the scalar curvature R of the metric g by the equation:

$$\Delta_g \phi + \frac{n-2}{4(n-1)} R\phi - \frac{1}{4(n-1)} \|\nabla \phi\|_g^2 = 0,$$

where Δ_g represents the Laplace-Beltrami operator associated with the metric g, n is the dimension of the underlying manifold, and $\|\nabla \phi\|_g^2$ represents the squared norm of the gradient of ϕ with respect to the metric g.

Analytical and Numerical Solutions

The Lichnerowicz equation presents a rich mathematical structure, allowing for both analytical and numerical solutions depending on the specific setup and boundary conditions. Analytical solutions can be obtained in certain symmetric cases or for special choices of R and n. However, in more general scenarios, numerical methods, such as finite difference or finite element methods, are employed to approximate solutions.

Initial Data in Relativity

Solving the Lichnerowicz equation is particularly significant in the context of relativity theory for determining initial data for the Einstein field equations. Given a three-dimensional Riemannian manifold and a prescribed set of geometrical constraints, the Lichnerowicz equation is utilized to find solutions that satisfy these constraints and represent physically meaningful initial data for the evolution of the gravitational field.

Applications in Cosmology

The Lichnerowicz equation finds application in studying various phenomena in cosmology, where it provides insights into the evolution of the universe and the behavior of gravitational waves. It is used, for example, to analyze the initial stages of the universe's expansion and the formation of structures, as well as to investigate the propagation of gravitational waves emitted by astrophysical sources.

The Lichnerowicz equation stands as a fundamental equation in relativity theory and plays a crucial role in understanding the geometry of spacetime and the behavior of gravitational fields. By examining conformal metrics and their relationship to the scalar curvature, the equation provides valuable insights into the dynamics of the universe and contributes to the accurate modeling of gravitational phenomena in both analytical and numerical settings.

Python Code Snippet

Below is a Python code snippet that implements the Lichnerowicz equation along with the techniques for analyzing conformal metrics and determining initial data in relativity.

```python
import numpy as np
from scipy.sparse import diags
from scipy.sparse.linalg import spsolve

def laplace_beltrami_operator(n, metric):
    '''
    Construct the Laplace-Beltrami operator given a metric on a 1D
    ↪ grid.
    :param n: Number of points in the grid.
    :param metric: The conformal metric as a function of the grid
    ↪ points.
    :return: Laplace-Beltrami operator as a sparse matrix.
    '''
    main_diag = np.zeros(n)
    off_diag = np.zeros(n-1)

    for i in range(n):
        main_diag[i] = (metric[i] + (metric[i-1] if i > 0 else 0) +
        ↪ (metric[i+1] if i < n-1 else 0)) / 3

    for i in range(n-1):
        off_diag[i] = - metric[i+1] / (metric[i] * (1/n))   #
        ↪ Simplified difference

    diagonals = [main_diag, off_diag, off_diag]
    return diags(diagonals, [0, -1, 1]).tocsc()


def lichnerowicz_equation(n, R, boundary_conditions):
    '''
    Solve the Lichnerowicz equation given scalar curvature R and
    ↪ boundary conditions.
    :param n: Number of points in the grid.
    :param R: Scalar curvature as a function on the grid.
    :param boundary_conditions: Boundary values for the solution.
    :return: Solution of the Lichnerowicz equation.
    '''
    # Define the conformal metric
    metric = np.ones(n)   # Can be replaced with actual metric values

    # Construct the Laplace-Beltrami operator
    L = laplace_beltrami_operator(n, metric)

    # Initial guess for the solution
    phi = np.zeros(n)
```

```
    # Set boundary conditions
    phi[0] = boundary_conditions[0]
    phi[-1] = boundary_conditions[1]

    # Solve the linear system
    rhs = -R * phi   # Right-hand side based on Lichnerowicz equation
     ↪ formulation
    phi = spsolve(L, rhs)

    return phi


# Example parameters and usage
n = 100   # Number of grid points
R = np.linspace(0, 1, n)   # Example scalar curvature
boundary_conditions = [1.0, 0.0]   # Boundary conditions for phi

# Solve the Lichnerowicz equation
solution = lichnerowicz_equation(n, R, boundary_conditions)

# Output the solution
print("Solution to the Lichnerowicz equation:", solution)
```

This code defines two functions:

- laplace_beltrami_operator constructs the Laplace-Beltrami operator based on a conformal metric defined on a 1D grid.
- lichnerowicz_equation solves the Lichnerowicz equation given a specific scalar curvature and boundary conditions.

The provided example demonstrates how to set up the problem, solve the Lichnerowicz equation, and print the results, showcasing the mathematical modeling in the context of relativity theory.

Multiple Choice Questions

1. What is the main purpose of the Lichnerowicz equation in relativity theory?

    (a) To model the motion of planets

    (b) To establish the relationship between matter and the curvature of spacetime

    (c) To determine the energy levels of atoms

    (d) To solve the heat equation

2. Which of the following quantities does the Lichnerowicz equation relate to the conformal factor ϕ?

    (a) The metric tensor g
    (b) The scalar curvature R
    (c) The gravitational constant G
    (d) The Riemann curvature tensor

3. In the context of the Lichnerowicz equation, what does the term $\Delta_g \phi$ represent?

    (a) The gradient of the conformal factor
    (b) The divergence of a vector field
    (c) The Laplace-Beltrami operator acting on ϕ
    (d) The area of the manifold

4. What is the significance of conformal metrics in general relativity?

    (a) They simplify the equations of motion for particles
    (b) They allow the study of spacetime through scale changes
    (c) They are solely used in cosmology
    (d) They are not relevant in relativity

5. Which of the following is true about the numerical methods applied to solve the Lichnerowicz equation?

    (a) They can only use the finite element method
    (b) They always yield exact solutions
    (c) They may involve finite difference or finite element methods
    (d) They are unnecessary since analytical solutions are always possible

6. In what context is the Lichnerowicz equation predominantly applied?

    (a) Quantum mechanics
    (b) Thermodynamics
    (c) Cosmology
    (d) Classical mechanics

7. The Lichnerowicz equation contributes to understanding gravitational fields primarily by:

   (a) Providing solutions for black hole dynamics
   (b) Establishing metrics for the vacuum
   (c) Examining the evolution of scalar fields
   (d) Analyzing the curvature of spacetime

Answers:
1. **B: To establish the relationship between matter and the curvature of spacetime** The Lichnerowicz equation arises from the need to relate geometric properties of spacetime (curvature) to the matter content as described by the Einstein field equations.

2. **B: The scalar curvature R** The Lichnerowicz equation connects the conformal factor ϕ with the scalar curvature R, indicating how the geometry of spacetime is deformed.

3. **C: The Laplace-Beltrami operator acting on ϕ** In the Lichnerowicz equation, $\Delta_g \phi$ indicates the Laplace-Beltrami operator's effect on the conformal factor, linking it to curvature dynamics.

4. **B: They allow the study of spacetime through scale changes** Conformal metrics help analyze the geometry of spacetime while preserving angles, thus facilitating studies that involve scaling without changing the fundamental structure.

5. **C: They may involve finite difference or finite element methods** Numerical solutions for the Lichnerowicz equation can use various methods, including finite differences and finite elements, depending on specific cases and boundary conditions.

6. **C: Cosmology** The Lichnerowicz equation is significant in cosmological studies, particularly in examining the expansion of the universe and gravitational phenomena over large scales.

7. **D: Analyzing the curvature of spacetime** The main contribution of the Lichnerowicz equation is its role in understanding the geometry of spacetime and how gravitational fields interact with that geometry, particularly in the context of initial data for the Einstein field equations.

Chapter 25

The Obstacle Problem

Variational Inequality Formulation

The obstacle problem is a classical problem in mathematical analysis and partial differential equations that arises in various fields, including physics, engineering, and finance. The problem involves finding the solution to a variational inequality defined over a given domain with prescribed boundary conditions.

In the obstacle problem, we seek a function u that satisfies the following variational inequality:

$$\text{Find } u \in K : \quad \int_\Omega \nabla u \cdot \nabla (v - u)\, dx \geq \int_\Omega f(v - u)\, dx \quad \forall v \in K,$$

where K is a closed set representing a constraint on the possible values of u, Ω is the domain of interest, and f is a given function.

Free Boundary Problems

An interesting aspect of the obstacle problem is the presence of free boundary conditions. The solution u is not known a priori on the entire boundary of Ω, but rather only on a subset Γ_u known as the free boundary. The free boundary is part of the boundary where the solution reaches the constraint set K and plays a crucial role in the behavior of the solution.

Regularity Theory

The obstacle problem exhibits rich mathematical structure and connections with various areas of analysis. Regularity theory for the obstacle problem is concerned with understanding the smoothness properties of the solution u and the free boundary Γ_u. This theory establishes conditions under which the solution is differentiable, allowing for further analysis of the problem.

Notably, the regularity theory of the obstacle problem is closely related to the theory of nonlinear potential theory, partial differential equations, and variational inequalities. It provides valuable insights into the behavior of the solution near the free boundary and contributes to the understanding of nonlinear phenomena in mathematical analysis.

Numerical Solutions

Numerical methods play a crucial role in solving the obstacle problem due to the complexity and nonlinearity of the variational inequality. Various numerical approaches have been developed, including finite difference methods, finite element methods, and variational approaches such as the penalty method and the augmented Lagrangian method. These methods allow for efficient and accurate approximation of the solution and the free boundary.

Unique challenges arise in numerical methods when dealing with the free boundary, as its location and shape are unknown and subject to change with each iteration of the numerical algorithm. Strategies such as adaptive mesh refinement, level set methods, and front tracking methods have been developed to address these challenges and provide reliable numerical solutions to the obstacle problem.

Applications in Finance and Physics

The obstacle problem finds applications in diverse fields, including finance and physics. In finance, the obstacle problem is used to model various pricing and optimization problems, such as the pricing of American options and the optimal portfolio selection in the presence of transaction costs. The obstacle problem formulation allows for the incorporation of constraints and nonlinearity inherent in these problems.

In physics, the obstacle problem arises in the study of equilibrium configurations of elastic bodies, where the solution corresponds to the displacement field of the body subject to given external forces and constraints. The obstacle problem formulation captures the non-penetration conditions and provides insights into the stability and behavior of the material.

The obstacle problem, with its variational inequality formulation and free boundary conditions, plays a fundamental role in mathematical analysis and its applications to various fields. From the regularity theory to numerical methods and real-world applications, the obstacle problem provides a valuable framework for understanding nonlinear phenomena and finding optimal solutions under constraints.

Python Code Snippet

Below is a Python code snippet that implements the variational inequality formulation of the obstacle problem, solves the free boundary using finite difference methods, and provides a numerical approximation of the solution.

```python
import numpy as np
import matplotlib.pyplot as plt

def initialize_mesh(domain, N):
    '''
    Initialize the mesh grid for the domain.
    :param domain: Tuple of (xmin, xmax, ymin, ymax).
    :param N: Number of grid points in each dimension.
    :return: Arrays of grid points.
    '''
    x = np.linspace(domain[0], domain[1], N)
    y = np.linspace(domain[2], domain[3], N)
    return np.meshgrid(x, y)

def solve_obstacle_problem(N, domain, f, g):
    '''
    Solve the obstacle problem using finite difference method.
    :param N: Number of grid points in each dimension.
    :param domain: Tuple of (xmin, xmax, ymin, ymax).
    :param f: Function representing the external force.
    :param g: Function representing the obstacle.
    :return: Numerical solution on the grid.
    '''
    x = np.linspace(domain[0], domain[1], N)
```

```python
    y = np.linspace(domain[2], domain[3], N)
    u = np.zeros((N, N))  # Initialize solution array

    dx = (domain[1] - domain[0]) / (N - 1)
    dy = (domain[3] - domain[2]) / (N - 1)

    for i in range(1, N-1):
        for j in range(1, N-1):
            # Finite difference approximation for Laplacian
            u_new = (u[i+1, j] + u[i-1, j] + u[i, j+1] + u[i, j-1])
            ↪  / 4
            u_new = max(u_new, g(x[i], y[j]))  # Apply obstacle
            ↪  condition
            u[i, j] = u_new

    return u, x, y

def external_force(x, y):
    '''
    Define the external force function f(x, y).
    :param x: x-coordinates.
    :param y: y-coordinates.
    :return: Value of external force.
    '''
    return np.sin(np.pi * x) * np.cos(np.pi * y)

def obstacle_function(x, y):
    '''
    Define the obstacle function g(x, y).
    :param x: x-coordinates.
    :param y: y-coordinates.
    :return: Value of the obstacle.
    '''
    return 0.5 * (x**2 + y**2)

# Parameters
N = 50  # Grid size
domain = (-1, 1, -1, 1)  # Domain in the form (xmin, xmax, ymin,
↪  ymax)

# Solve the obstacle problem
u, x, y = solve_obstacle_problem(N, domain, external_force,
↪  obstacle_function)

# Plot the results
plt.figure(figsize=(8, 6))
plt.contourf(x, y, u, levels=50, cmap='jet')
plt.colorbar(label='Solution u(x, y)')
plt.title('Solution to the Obstacle Problem')
plt.xlabel('x')
plt.ylabel('y')
plt.title('Numerical Solution of Obstacle Problem')
```

```
plt.show()
```

This code defines several functions:

- `initialize_mesh` sets up the grid for the computational domain.
- `solve_obstacle_problem` implements the finite difference method to solve the obstacle problem, applying the variational inequality based on the given external force and obstacle functions.
- `external_force` defines the external force $f(x, y)$ that influences the problem.
- `obstacle_function` specifies the obstacle $g(x, y)$ that constrains the solution.

The provided example computes the numerical solution of the obstacle problem using a finite difference method and visualizes the results with a contour plot.

Multiple Choice Questions

1. What type of problem does the obstacle problem involve?

    (a) A linear equation

    (b) A variational inequality

    (c) A stochastic process

    (d) A system of ordinary differential equations

2. In the variational inequality formulation of the obstacle problem, which space does the function u belong to?

    (a) The $L^2 space A closed convex set K$

    (b) The continuous function space

    (d) The L^∞ space

3. What characterizes the free boundary Γ_u in the obstacle problem?

    (a) It is a boundary where the solution value is always zero.

    (b) It defines where the solution touches the constraint set K.

    (c) It separates the domain from a non-physical region.

    (d) It is a fixed boundary throughout the solution process.

4. Which numerical method is commonly used for solving the obstacle problem?

    (a) Finite difference method

    (b) Lattice Boltzmann method

    (c) Double precision arithmetic

    (d) Reinforcement learning methods

5. The regularity theory of the obstacle problem is particularly concerned with:

    (a) The well-posedness of linear equations.

    (b) The smoothness of the solution and its free boundary.

    (c) The dimensionality of the solution space.

    (d) The conservation laws governing the dynamics.

6. In which field is the obstacle problem notably applied?

    (a) Number theory

    (b) Quantum mechanics

    (c) Financial derivatives pricing

    (d) Topology

7. Which method helps in approximating the location of the free boundary?

    (a) Adaptive mesh refinement

    (b) Gradient descent

    (c) Simplex method

    (d) Newton's method

Answers:

1. **B: A variational inequality** The obstacle problem is classically defined as a variational inequality, where the solution is constrained by an obstacle or boundary condition.

2. **B: A closed convex set K** The function u in the obstacle problem belongs to a closed set K, which defines the constraints for the solution.

3. **B: It defines where the solution touches the constraint set K.** The free boundary Γ_u characterizes the points in

the domain where the solution meets the constraint set, making it essential for analyzing the solution's properties.

4. **A: Finite difference method** Finite difference methods are frequently employed to numerically solve the obstacle problem due to their effectiveness in approximating solutions of partial differential equations.

5. **B: The smoothness of the solution and its free boundary.** Regularity theory focuses on understanding how smooth or differentiable the solution and its associated free boundary are, which is crucial in the analysis of the obstacle problem.

6. **C: Financial derivatives pricing** The obstacle problem has significant applications in finance, particularly in the pricing of American options, where constraints affect the pricing strategies.

7. **A: Adaptive mesh refinement** Adaptive mesh refinement is a numerical strategy used to improve the accuracy of approximating the solution and capturing the dynamics of the free boundary effectively.

Chapter 26

The Tolman-Oppenheimer-Volkoff Equation

Astrophysical Context

The Tolman-Oppenheimer-Volkoff (TOV) equation is a fundamental equation in astrophysics that describes the structure and properties of dense, compact objects such as neutron stars. These objects, formed from the remnants of massive stars, are characterized by high densities and intense gravitational fields.

Mathematical Derivation

The TOV equation is derived by considering the equilibrium of a spherically symmetric, static, and isotropic object subject to gravitational forces. By applying the principles of general relativity, the gravitational effects are described by the Einstein field equations, which provide a set of differential equations relating the geometry of spacetime to the distribution of matter and energy.

Applying the assumptions of spherical symmetry and hydrostatic equilibrium, the TOV equation is obtained by balancing the pressure gradient force and the gravitational force. It takes the form:

$$\frac{dP}{dr} = -\frac{G\left(\varepsilon + P\right)\left(m + 4\pi r^3 P\right)}{c^2 r \left(r - 2Gm/c^2\right)},$$

where $P(r)$ is the pressure as a function of radial distance r, ε is the energy density, $m(r)$ is the mass enclosed within radius r, G is the gravitational constant, and c is the speed of light.

Analytical Solutions

Exact analytical solutions of the TOV equation are extremely challenging to obtain due to its nonlinearity and complexity. However, simplified models and asymptotic approximations have been developed to study specific cases and gain insights into the behavior of neutron stars.

By assuming specific equations of state relating the pressure, energy density, and composition of matter, approximate solutions can be obtained. Notable examples include the polytropic equation of state, which assumes a power-law relationship between pressure and energy density, and the ideal Fermi gas equation of state, which considers degenerate matter composed of fermions.

Numerical Methods

Due to the difficulty of obtaining analytical solutions, numerical methods play a crucial role in solving the TOV equation and understanding neutron star properties. Finite difference methods, spectral methods, and numerical relativity techniques are commonly employed to solve the system of coupled ordinary differential equations represented by the TOV equation.

The numerical solution of the TOV equation requires the specification of an equation of state and appropriate boundary conditions. The solution provides insights into the mass-radius relationship, the central density and pressure, and the maximum mass of neutron stars, which is of great interest in astrophysics.

Applications in Stellar Structure

The TOV equation and its solutions have provided significant contributions to our understanding of the internal structure and properties of neutron stars. By considering various equations of state, the equation allows for the study of different constituents of neutron star matter, such as nuclear matter, hyperons, and exotic forms of matter like quark matter.

The TOV equation also plays a crucial role in determining the maximum possible mass of neutron stars, known as the Tolman-Oppenheimer-Volkoff limit. By considering different equations of state and symmetry energy effects, constraints can be placed on the properties of dense matter and help refine our understanding of the nature of extreme astrophysical objects.

The TOV equation, with its mathematical derivation and numerical solution techniques, provides a fundamental framework for studying neutron stars and exploring the limits of dense matter in astrophysics. Its applications go beyond astrophysics, as similar equations arise in diverse fields such as condensed matter physics and general relativity.

Python Code Snippet

Below is a Python code snippet that implements the Tolman-Oppenheimer-Volkoff (TOV) equation and solves it numerically to analyze the structure of neutron stars.

```python
import numpy as np
import matplotlib.pyplot as plt
from scipy.integrate import odeint

def tov_equations(y, r, G, c):
    '''
    Differential equations for the TOV equation.
    :param y: Array containing the values [m, P] where m is mass and
    ↪ P is pressure.
    :param r: Radial distance.
    :param G: Gravitational constant.
    :param c: Speed of light.
    :return: Derivative of m and P.
    '''
    m, P = y
     = equation_of_state(P)  # Energy density derived from the
    ↪ equation of state
    dP_dr = -G * ( + P) * (m + 4 * np.pi * r**3 * P) / (c**2 * r *
    ↪ (r - 2 * G * m / c**2))
    dm_dr = 4 * np.pi * r**2 * 
    return [dm_dr, dP_dr]

def equation_of_state(P):
    '''
```

```
    Example equation of state for a polytropic model: P = K * ^(1 +
 ↪  1/n)
    :param P: Pressure.
    :return: Energy density.
    '''
    n = 1    # Polytropic index
    K = 1.0  # Polytropic constant
     = K * (P ** (n / (n + 1)))
    return

def solve_tov(r_max, G, c, P_c):
    '''
    Solves the TOV equations using odeint.
    :param r_max: Maximum radius to calculate.
    :param G: Gravitational constant.
    :param c: Speed of light.
    :param P_c: Central pressure.
    :return: Radial distances, masses, and pressures.
    '''
    r = np.linspace(1e-10, r_max, 1000)  # Avoid division by zero at
 ↪  r=0
    y0 = [0, P_c]  # Initial conditions: m = 0 at r = 0, central
 ↪  pressure P_c

    sol = odeint(tov_equations, y0, r, args=(G, c))
    return r, sol[:, 0], sol[:, 1]

# Constants
G = 6.67430e-11   # Gravitational constant (m^3 kg^-1 s^-2)
c = 299792458     # Speed of light (m/s)
P_c = 1e34        # Central pressure in Pascals

# Solve the TOV equation
r_max = 15e3  # Maximum radius in meters
r, m, P = solve_tov(r_max, G, c, P_c)

# Plotting the results
plt.figure(figsize=(12, 6))
plt.subplot(1, 2, 1)
plt.plot(r, m, label='Mass (kg)')
plt.xlabel('Radius (m)')
plt.ylabel('Mass')
plt.title('Mass vs Radius')
plt.grid()
plt.legend()

plt.subplot(1, 2, 2)
plt.plot(r, P, label='Pressure (Pa)', color='r')
plt.xlabel('Radius (m)')
plt.ylabel('Pressure')
plt.title('Pressure vs Radius')
plt.grid()
plt.legend()
```

```
plt.tight_layout()
plt.show()
```

This code defines the following functions:

- `tov_equations` computes the derivatives of mass and pressure at a given radius based on the TOV equations.
- `equation_of_state` provides a simple polytropic equation of state relating pressure and energy density.
- `solve_tov` integrates the TOV equations to obtain mass and pressure profiles for a neutron star.

The provided example sets physical constants, initializes the central pressure, solves the TOV equations, and plots the mass and pressure as functions of the radius, thus providing insights into neutron star structure based on the derived TOV equation.

Multiple Choice Questions

1. The Tolman-Oppenheimer-Volkoff (TOV) equation primarily relates to which type of astronomical object?

    (a) Black holes

    (b) Main-sequence stars

    (c) Neutron stars

    (d) Red giants

2. What type of symmetry is assumed in deriving the TOV equation?

    (a) Cylindrical symmetry

    (b) Spherical symmetry

    (c) Asymmetry

    (d) Planar symmetry

3. Which of the following physical quantities is NOT included in the TOV equation?

    (a) Pressure P

    (b) Energy density ε

    (c) Temperature T

(d) Mass m

4. The term $(\varepsilon + P)$ in the TOV equation contributes to which of the following effects?

    (a) Gravitational binding energy

    (b) Pressure support against gravitational collapse

    (c) Thermal energy exchange

    (d) Electromagnetic force balance

5. Which of the following numerical methods is commonly utilized to solve the TOV equation?

    (a) Perturbation methods

    (b) Finite difference methods

    (c) Variational methods

    (d) Monte Carlo simulations

6. The maximum mass of neutron stars determined from the TOV equation is known as:

    (a) Chandrasekhar limit

    (b) Tolman-Oppenheimer-Volkoff limit

    (c) Schwarzschild limit

    (d) Planck limit

7. Which equation of state (EOS) assumes a relation of the form $P = K\varepsilon^{\Gamma}$, where K and Γ are constants?

    (a) Ideal gas law

    (b) Timelike EOS

    (c) Polytropic EOS

    (d) Fermi gas EOS

Answers:

1. **C: Neutron stars** The TOV equation is specifically formulated to describe the structure and stability of neutron stars, which are dense and compact stellar remnants.

2. **B: Spherical symmetry** The derivation of the TOV equation assumes that the object in question possesses spherical symmetry, allowing the use of general relativity principles to describe equilibrium.

3. **C: Temperature** T While pressure P, energy density ε, and mass m are included in the TOV equation, temperature is not typically considered in the basic static framework of the equation.

4. **B: Pressure support against gravitational collapse** The term $(\varepsilon + P)$ reflects the contribution of both energy density and pressure to counteract gravitational force, providing stability against collapse.

5. **B: Finite difference methods** Finite difference methods are a common numerical technique used to solve the differential equations derived from the TOV equation, allowing for computational modeling of neutron stars.

6. **B: Tolman-Oppenheimer-Volkoff limit** The maximum mass of neutron stars derived from the TOV equation is referred to as the Tolman-Oppenheimer-Volkoff limit, which sets constraints on the properties of dense matter.

7. **C: Polytropic EOS** The polytropic equation of state exhibits a power-law relationship between pressure and energy density, commonly used in astrophysics to model neutron star matter and simple stellar structures.

Chapter 27

The Yamabe Equation

Conformal Geometry Background

In this chapter, we explore the mathematical formulation and properties of the Yamabe equation, which arises in conformal geometry. Conformal geometry studies the geometric properties preserved under conformal transformations, which are transformations that locally preserve angles but permit changes in distances.

1 Conformal Transformations

A conformal transformation in two dimensions is a smooth mapping $f : M \to N$ between two Riemannian manifolds (M, g) and (N, h) that preserves angles. Specifically, for any two vectors v, w at a point p in M, the inner product $g_p(v, w)$ is proportional to the inner product $h_{f(p)}(df(v), df(w))$ at the corresponding point $f(p)$ in N. Mathematically, this can be expressed as $h_{f(p)}(df(v), df(w)) = \lambda(p) \cdot g_p(v, w)$, where $\lambda(p)$ is the conformal factor.

2 Yamabe Flow

The Yamabe flow is a process that transforms the metric of a manifold to achieve constant scalar curvature. The motivation behind this flow comes from the Yamabe problem, which asks for a conformal metric with constant scalar curvature.

The Yamabe equation arises as a critical point of the total scalar curvature functional. Given a smooth, compact Riemannian manifold (M, g), the Yamabe equation is given by:

$$P_g(u) = -\Delta_g u + \frac{n}{2(n-1)} R_g u + K_g u^{\frac{n+2}{n-2}} = 0$$

where Δ_g is the Laplace-Beltrami operator, R_g is the scalar curvature, K_g is the Gaussian curvature, and n is the dimension of the manifold.

3 Existence and Uniqueness

The existence and uniqueness of solutions to the Yamabe equation depend on the geometric properties of the manifold and the prescribed boundary conditions. In the case of a closed manifold with constant scalar curvature, solutions exist and are unique up to scaling. However, for general manifolds, the existence and uniqueness of solutions are not guaranteed.

4 Conformal Deformation and Positive Yamabe Constant

The Yamabe constant is a fundamental quantity that characterizes the conformal geometry of a manifold. It measures the lowest scalar curvature achieved under conformal deformations of the metric.

For a compact manifold (M, g), the Yamabe constant $Y(M)$ is defined as:

$$Y(M) = \inf\{\sigma(g) : g \text{ is a smooth metric on } M\}$$

where $\sigma(g)$ is the total scalar curvature, given by $\sigma(g) = \int_M R_g dV_g$.

The positivity of the Yamabe constant plays a crucial role in the study of the Yamabe equation. When the Yamabe constant is positive, it guarantees the existence of a solution to the Yamabe equation.

Mathematical Formulation

In this section, we present the mathematical formulation of the Yamabe equation and discuss its key properties. The Yamabe equation is a nonlinear, partial differential equation that involves the Laplace-Beltrami operator and the scalar curvature of a Riemannian manifold.

Given a smooth, compact Riemannian manifold (M, g) of dimension n, the Yamabe equation is given by:

$$P_g(u) = -\Delta_g u + \frac{n}{2(n-1)} R_g u + K_g u^{\frac{n+2}{n-2}} = 0$$

where Δ_g is the Laplace-Beltrami operator, R_g is the scalar curvature, K_g is the Gaussian curvature, and u is a smooth function defined on M.

1 Yamabe Conformal Class

The Yamabe equation is inherently related to conformal deformations of the metric. In particular, it seeks a metric $\tilde{g}$ in the same conformal class as g that achieves constant scalar curvature.

The conformal class of a Riemannian metric g on a manifold M is defined as the set of all metrics $\tilde{g}$ on M that are conformally equivalent to g. Mathematically, the conformal class of g is denoted as $[g]$ and is defined as:

$$[g] = \{\tilde{g} = e^{2u} g : u \text{ is a smooth function on } M\}$$

The Yamabe conformal class is then defined as the conformal class of g that achieves constant scalar curvature. It is denoted as $[g]_{\text{Yamabe}}$ and is given by

$$[g]_{\text{Yamabe}} = \{\tilde{g} : \tilde{g} \text{ has constant scalar curvature}\}$$

2 Yamabe Problem

The Yamabe problem seeks a conformal metric in the Yamabe conformal class that achieves constant scalar curvature. In other words, the problem aims to find a metric $\tilde{g}$ in the conformal class $[g]_{\text{Yamabe}}$ such that $\tilde{g}$ has constant scalar curvature.

The Yamabe problem can be stated mathematically as follows: Given a smooth, compact Riemannian manifold (M, g) of dimension n, find a metric $\tilde{g}$ in the conformal class $[g]_{\text{Yamabe}}$ such that $\tilde{g}$ has constant scalar curvature.

Solutions to the Yamabe problem are closely related to solutions of the Yamabe equation. Specifically, a metric $\tilde{g}$ in the Yamabe conformal class $[g]_{\text{Yamabe}}$ achieves constant scalar curvature if and only if the associated smooth function u satisfies the Yamabe equation.

Existence and Uniqueness

The existence and uniqueness of solutions to the Yamabe equation heavily rely on the geometric properties of the underlying manifold, as well as the prescribed boundary conditions.

1 Closed Manifolds with Constant Scalar Curvature

For closed manifolds (compact manifolds without boundary) with constant scalar curvature, solutions to the Yamabe equation always exist and are unique up to scaling. This result is known as the Yamabe problem on closed manifolds.

Specifically, if the manifold (M, g) is closed and has constant scalar curvature, then there exists a smooth function u such that the conformally related metric $\tilde{g} = u^{-\frac{4}{n-2}} g$ has constant scalar curvature.

2 General Manifolds and Boundary Conditions

For general manifolds, the existence and uniqueness of solutions to the Yamabe equation are not guaranteed. The presence of boundary conditions and geometric constraints can significantly influence the solutions.

Various geometric conditions and restrictions, such as the prescribing function method and the admissible Yamabe class, have been studied to establish sufficient conditions for the existence and uniqueness of solutions. These conditions often involve the geometry of the manifold and the prescribed boundary conditions.

The study of existence and uniqueness results for the Yamabe equation remains an active area of research, and further advancements continue to deepen our understanding of this fundamental problem in conformal geometry.

Numerical Methods

Due to the complexity of the Yamabe equation and its nonlinear nature, numerical methods play a crucial role in obtaining approximate solutions and analyzing the behavior of solutions.

1 Finite Difference Methods

Finite difference methods are widely used to discretize the Yamabe equation and solve it numerically. The continuous differential equation is approximated on a discretized grid, and finite difference approximations are used to discretize the Laplace-Beltrami operator and the scalar curvature term. The resulting system of algebraic equations can then be solved using numerical linear algebra techniques.

2 Variational Methods

Variational methods are another important approach employed to study the Yamabe equation numerically. By formulating the Yamabe equation as a variational problem, the solutions can be obtained by minimizing a suitable functional over a set of admissible functions.

These numerical methods allow for the investigation of specific instances of the Yamabe equation, such as finding solutions for a given geometric configuration or prescribed boundary conditions. They provide insights into the behavior and properties of solutions in a computational framework.

Applications in Mathematical Physics

The Yamabe equation has far-reaching applications beyond the realm of conformal geometry. The solutions to the Yamabe equation have connections to various areas of mathematical physics, where conformal transformations and curvature play fundamental roles.

1 Quantum Field Theory

In quantum field theory, the conformal symmetry is a fundamental symmetry concept that underlies the behavior of elementary particles and their interactions. The solutions to the Yamabe equation help characterize the conformal structure of spacetime in the context of quantum field theory.

2 String Theory

String theory, a theoretical framework that aims to unify quantum mechanics and general relativity, heavily relies on conformal geometry. The solutions to the Yamabe equation provide important insights into the conformal structure of spacetime in the context of string theory.

3 General Relativity

The Yamabe equation has connections to general relativity, the theory of gravity formulated by Einstein. The conformal transformations captured by the Yamabe equation are relevant in understanding the geometric properties of spacetime and the behavior of gravitational fields.

4 Geometric Analysis

The Yamabe equation has broad implications in the field of geometric analysis. It reveals deep connections between various geometric quantities, such as scalar curvature and conformal deformations, providing insights into the intrinsic geometry of manifolds.

5 Quantum Field Theory

6 Functional Analysis

The Yamabe equation's applications in mathematical physics highlight its importance in understanding fundamental concepts and structures in theoretical physics. The deep interplay between conformal geometry, curvature, and physical phenomena continues to inspire new developments in mathematical physics.

Conclusion

In this chapter, we have explored the mathematical formulation, properties, and numerical approaches related to the Yamabe equation. We discussed its connection to conformal geometry, the existence and uniqueness of solutions, and its applications in mathematical physics. The Yamabe equation serves as a powerful tool to investigate the geometric properties of manifolds and reveals crucial insights into various areas of mathematics and theoretical physics. Through ongoing research, further advancements will continue to deepen our understanding of the Yamabe equation and its implications in diverse fields.

Python Code Snippet

Below is a Python code snippet that solves the Yamabe equation defined in this chapter using numerical methods based on finite differences. The code includes functions for discretizing the given equation, solving it using a simple iterative method, and visualizing the results.

```python
import numpy as np
import matplotlib.pyplot as plt

def laplace_beltrami(u, dx):
    '''
    Compute the Laplace-Beltrami operator for a given function u
    ↪ using finite differences.
    :param u: The function values on the grid.
    :param dx: The grid spacing.
    :return: The Laplace-Beltrami operator applied to u.
    '''
    # Applying finite differences to compute the Laplacian
    lap_u = np.zeros_like(u)
    lap_u[1:-1] = (u[:-2] - 2 * u[1:-1] + u[2:]) / dx**2
    return lap_u

def yamabe_equation_solver(n, num_iterations, R, K, dt):
    '''
    Solves the Yamabe equation using an iterative method.
    :param n: Number of grid points.
    :param num_iterations: Number of iterations for convergence.
    :param R: Scalar curvature.
    :param K: Gaussian curvature.
    :param dt: Time step for iterative updating.
    :return: Approximate solution u at the grid points.
    '''
    # Initialize the grid and function u
    x = np.linspace(0, 1, n)
    u = np.random.rand(n)   # Initial guess

    dx = x[1] - x[0]   # Grid spacing
    for _ in range(num_iterations):
        lap_u = laplace_beltrami(u, dx)
        # Update u based on the Yamabe equation
        u[1:-1] += dt * (-lap_u[1:-1] + (n/2/(n-1)) * R * u[1:-1] +
        ↪      K * u[1:-1]**((n+2)/(n-2)))
        # Normalize the solution to prevent explosion
```

```
            u /= np.max(np.abs(u))

        return u

def plot_solution(u):
    '''
    Plot the solution to the Yamabe equation.
    :param u: The function values to plot.
    '''
    plt.figure(figsize=(8, 4))
    plt.plot(u, label='Approximate Solution u')
    plt.title('Solution to the Yamabe Equation')
    plt.xlabel('Grid Points')
    plt.ylabel('Function Value')
    plt.legend()
    plt.grid()
    plt.show()

# Parameters for the Yamabe equation
n = 100           # Number of grid points
num_iterations = 500  # Number of iterations for convergence
R = 1.0           # Scalar curvature
K = 1.0           # Gaussian curvature
dt = 0.01         # Time step for iterative updating

# Solve the Yamabe equation and plot the results
u = yamabe_equation_solver(n, num_iterations, R, K, dt)
plot_solution(u)
```

This code defines three main functions:

- `laplace_beltrami` computes the Laplace-Beltrami operator for a given function using finite differences.
- `yamabe_equation_solver` iteratively solves the Yamabe equation by updating the function based on the equation's specifications.
- `plot_solution` visualizes the approximate solution of the Yamabe equation.

The provided example sets up the necessary parameters for the problem, runs the solver, and plots the resulting function, allowing for an insightful analysis of the Yamabe equation's behavior in a numerical framework.

Multiple Choice Questions

1. Which mathematical operator is utilized in the formulation of the Yamabe equation?

(a) The Laplace operator

(b) The Laplace-Beltrami operator

(c) The Wave operator

(d) The D'Alembert operator

2. The Yamabe constant $Y(M)$ is defined as:

    (a) The maximum scalar curvature attained on the manifold M

    (b) The infimum of the total scalar curvature over all metrics in the Yamabe class

    (c) The average scalar curvature over the manifold M

    (d) The total volume of the manifold M

3. In the context of the Yamabe problem, what does it seek to find?

    (a) A metric with maximum scalar curvature

    (b) A conformal metric with constant scalar curvature

    (c) A metric that minimizes the volume of the manifold

    (d) A smooth function that does not vary across the manifold

4. The Yamabe flow ultimately aims to achieve what for the metric?

    (a) Unbounded curvature

    (b) Constant scalar curvature

    (c) Varied curvature

    (d) Zero scalar curvature

5. For which type of manifolds does the Yamabe equation guarantee existence and uniqueness of solutions?

    (a) Open manifolds

    (b) Non-compact manifolds

    (c) Closed manifolds with constant scalar curvature

    (d) Only flat manifolds

6. Which of the following areas has recently shown a significant interest in the applications of the Yamabe equation?

(a) Number theory

(b) Combinatorial optimization

(c) Quantum field theory

(d) Game theory

7. In numerical methods for solving the Yamabe equation, which method transforms the computational problem into minimizing a functional over admissible functions?

(a) Finite difference methods

(b) Variational methods

(c) Spectral methods

(d) Direct simulation

Answers:
1. **B: The Laplace-Beltrami operator** The Yamabe equation is characterized by the use of the Laplace-Beltrami operator, which generalizes the Laplacian to Riemannian manifolds and is crucial in the study of geometric and analytical properties of these structures.

2. **B: The infimum of the total scalar curvature over all metrics in the Yamabe class** The Yamabe constant $Y(M)$ is defined as the infimum of the total scalar curvature achieved under conformal deformations of the metric on a compact manifold (M, g).

3. **B: A conformal metric with constant scalar curvature** The Yamabe problem specifically seeks to find such a metric within the conformal class of the given metric g on the compact manifold.

4. **B: Constant scalar curvature** The aim of the Yamabe flow is to adjust the metric on the manifold so that the resulting metric displays constant scalar curvature, which is significant for various geometric interpretations.

5. **C: Closed manifolds with constant scalar curvature** The existence and uniqueness theorem for the Yamabe equation guarantees solutions specifically for closed (compact without boundary) manifolds characterized by constant scalar curvature.

6. **C: Quantum field theory** The solutions to the Yamabe equation have implications and applications in quantum field theory, particularly concerning the conformal structures of spacetime as explored in theoretical physics.

7. **B: Variational methods** Variational methods involve formulating the Yamabe equation as a variational problem, enabling numerical approximations through minimization of a functional associated with the equation.

Chapter 28

The Dirichlet Problem

Boundary Value Problem Setup

In this chapter, we explore the mathematical formulation and properties of the Dirichlet problem, a fundamental boundary value problem in the theory of partial differential equations (PDEs). The Dirichlet problem deals with finding solutions to a given PDE that satisfy prescribed boundary conditions on the boundary of a domain.

1 Dirichlet Boundary Conditions

The Dirichlet boundary conditions specify the values of the solution on the boundary of a domain. Specifically, for a given PDE defined on a domain D with boundary ∂D, the Dirichlet boundary conditions prescribe the values of the solution u on ∂D.

Mathematically, the Dirichlet boundary conditions can be expressed as:

$$u(x) = g(x) \quad \text{for } x \in \partial D$$

where g is a given function representing the boundary values of the solution.

2 The Dirichlet Problem

The Dirichlet problem aims to find a solution to a given PDE subject to the specified Dirichlet boundary conditions. In other words, it seeks a function u that satisfies both the PDE and the prescribed boundary values.

The Dirichlet problem can be stated mathematically as follows: Given a PDE in a domain D with boundary ∂D and a specified function g on ∂D, find a solution u that satisfies the PDE and the Dirichlet boundary conditions.

Solving the Dirichlet problem is crucial in many areas of mathematics and engineering, as it allows us to determine the behavior of solutions to PDEs in a given domain with specified boundary values.

Weak Solutions and Variational Formulation

In this section, we explore the concept of weak solutions and the variational formulation of the Dirichlet problem. Weak solutions provide a broader framework for solving PDEs that may not have classical solutions.

1 Sobolev Spaces

To define weak solutions, we need to introduce the concept of Sobolev spaces. Sobolev spaces consist of functions that possess sufficient weak derivatives, allowing us to generalize the notion of differentiation for functions that may not be smooth.

Denoted as $H^1(D)$, the Sobolev space $H^1(D)$ consists of functions defined on the domain D whose first-order weak derivatives are square-integrable. Mathematically, a function u belongs to $H^1(D)$ if its distributional derivative exists and the integral of the square of its weak derivative is finite:

$$\|u\|^2_{H^1(D)} = \|u\|^2_{L^2(D)} + \|\nabla u\|^2_{L^2(D)} < \infty$$

where ∇u represents the weak gradient of u.

2 Weak Solutions

A weak solution to the Dirichlet problem is a function that satisfies the PDE in a weak sense. Instead of requiring the solution to satisfy the PDE pointwise, weak solutions are defined through the concept of distributional derivatives and test functions.

For a given PDE and Dirichlet boundary conditions, a function u is a weak solution if it belongs to the appropriate Sobolev space and satisfies the weak formulation of the PDE.

3 Variational Formulation

The variational formulation provides an alternative approach to the weak solution of the Dirichlet problem. It formulates the problem as a minimization problem involving a suitable functional.

In the variational formulation, the Dirichlet problem is equivalent to finding the critical points of a functional called the Dirichlet energy. The Dirichlet energy, denoted as $E(u)$, is defined as:

$$E(u) = \frac{1}{2}\int_D |\nabla u|^2 - \int_D fu$$

where f represents the source term in the PDE.

The variational formulation seeks a function u that minimizes the Dirichlet energy among all functions in a certain function space that satisfy the Dirichlet boundary conditions. The critical points of E correspond to weak solutions of the Dirichlet problem.

Classical Solutions

In this section, we explore the properties of classical solutions to the Dirichlet problem. Classical solutions are typically smooth solutions that satisfy the PDE exactly.

1 Finite Difference Methods

Finite difference methods are commonly used to approximate classical solutions to the Dirichlet problem. These methods discretize the domain and replace derivatives with difference quotients, allowing for easier calculation of approximate solutions on a grid.

By applying finite difference approximations to the PDE and boundary conditions, a system of algebraic equations can be obtained, which can then be solved using numerical linear algebra techniques.

2 Finite Element Methods

Finite element methods provide another powerful approach for finding classical solutions to the Dirichlet problem. These methods divide the domain into smaller, simpler subdomains called elements.

Within each element, the solution is represented by piecewise polynomial functions, and the PDE is approximated by enforcing the weak form within each element. The resulting system of equations can be solved to obtain the values of the solution at each node or vertex.

Finite element methods offer flexibility in handling complex geometries and allow for adaptive mesh refinement, making them well-suited for problems with irregular boundaries or varying material properties.

Numerical Methods

Numerical methods play a crucial role in solving the Dirichlet problem. They allow us to approximate solutions for domains where analytical or classical solutions are difficult or impossible to obtain.

1 Finite Difference Methods

Finite difference methods discretize the domain and replace derivatives with difference quotients, resulting in a system of algebraic equations. These equations can be solved using standard numerical techniques, such as iterative methods or direct methods.

Using finite difference methods, the Dirichlet problem can be approximated on a grid, enabling the computation of numerical solutions. The accuracy of the approximation depends on the grid spacing and the order of the difference approximations.

2 Finite Element Methods

Finite element methods provide a powerful framework for solving the Dirichlet problem numerically. These methods discretize the domain into simpler elements and represent the solution as a combination of piecewise polynomial functions within each element.

By formulating the problem in terms of a variational principle, finite element methods convert the PDE into a system of algebraic equations. This system can be solved using numerical linear algebra techniques, such as Gaussian elimination or iterative methods.

Finite element methods offer flexibility in handling complex geometries and allow for adaptive mesh refinement, which can significantly enhance the accuracy of numerical solutions.

Applications in Engineering

The Dirichlet problem finds numerous applications in engineering, as it provides a means to determine the behavior of solutions to PDEs in various engineering domains. Some notable applications include:

1 Structural Analysis

In structural analysis, the Dirichlet problem helps determine the deformation and stress distribution in solid structures subject to prescribed boundary displacements or forces.

By formulating the governing PDEs and boundary conditions, structural engineers can solve the corresponding Dirichlet problem to obtain solutions that describe the behavior of structures under different loading conditions. This information is crucial for designing safe and efficient structures.

2 Heat Transfer

In heat transfer analysis, the Dirichlet problem is instrumental in solving the heat conduction equation with prescribed temperature conditions on the boundaries of a domain.

By solving the Dirichlet problem, engineers can determine the temperature distribution within a system, which is essential for designing efficient heat exchangers, thermal insulation materials, and other heat transfer devices.

3 Fluid Mechanics

In fluid mechanics, the Dirichlet problem helps determine the velocity field and pressure distribution in fluid flow problems subject to specified boundary conditions.

By solving the Navier-Stokes equations or other relevant PDEs in conjunction with the Dirichlet problem, engineers can obtain solutions that describe the flow behavior and pressure distribution in pipes, channels, and other fluid-carrying systems. This information is vital for designing efficient and reliable fluid transport systems.

4 Electromagnetics

In electromagnetics, the Dirichlet problem is used to determine the electric field or magnetic field distribution subject to prescribed boundary conditions.

By solving the relevant PDEs and the associated Dirichlet problem, engineers can analyze the behavior of electric or magnetic fields in different devices, such as antennas, electromagnetic waveguides, and circuits. This knowledge is essential for designing and optimizing electromagnetic systems.

5 Numerical Simulations

The Dirichlet problem is also crucial in performing numerical simulations for various engineering applications. By discretizing the PDE and applying numerical methods, engineers can simulate and predict the behavior of complex systems, leading to better design decisions and improved performance.

Numerical simulations based on solving the Dirichlet problem allow engineers to analyze and optimize the behavior of engineering systems in a cost-effective and efficient manner.

Conclusion

In this chapter, we have explored the mathematical formulation, properties, and numerical methods associated with the Dirichlet problem. The Dirichlet problem plays a central role in PDE theory and finds numerous applications in engineering, such as structural analysis, heat transfer, fluid mechanics, electromagnetics, and numerical simulations.

By formulating the problem with appropriate boundary conditions and applying numerical methods, engineers can determine the behavior of solutions to PDEs and design efficient and reliable systems. The study and solution of the Dirichlet problem continue to be an active area of research, driving advancements in engineering analysis and computational methods.Certainly! Below is a Python code snippet that includes the essential equations, formulas, and algorithms related to the Dirichlet problem and numerical methods discussed in the chapter.

Python Code Snippet

Below is a Python code snippet that implements the numerical solution of the Dirichlet problem using both finite difference and finite element methods. This code solves the Poisson equation in a 2D domain.

```python
import numpy as np
import matplotlib.pyplot as plt
from scipy.sparse import diags
from scipy.sparse.linalg import spsolve

def solve_poisson_fd(grid_size, boundary_conditions, source_term):
    '''
    Solves the Poisson equation using finite difference method.

    :param grid_size: Size of the grid (Nx, Ny).
    :param boundary_conditions: Boundary values for Dirichlet
    ↪ conditions.
    :param source_term: Source term defined on the grid.
    :return: Solution array.
    '''
    Nx, Ny = grid_size
    dx = 1.0 / (Nx + 1)
    dy = 1.0 / (Ny + 1)

    # Initialize the solution array
    u = np.zeros((Nx + 2, Ny + 2))

    # Apply boundary conditions
    u[0, :] = boundary_conditions['top']
    u[Nx + 1, :] = boundary_conditions['bottom']
    u[:, 0] = boundary_conditions['left']
    u[:, Ny + 1] = boundary_conditions['right']

    # Iterative method
    for it in range(10000):
        u_old = u.copy()
        # Update interior points
        for i in range(1, Nx + 1):
            for j in range(1, Ny + 1):
                u[i, j] = 0.25 * (u_old[i + 1, j] + u_old[i - 1, j]
                ↪ +
                                  u_old[i, j + 1] + u_old[i, j - 1]
                                  ↪ -
                                  dx ** 2 * source_term[i - 1, j -
                                  ↪ 1])

    return u

def plot_solution(u, title):
```

```
'''
Plots the solution to the Poisson equation.

:param u: Solution array.
:param title: Title of the plot.
'''
plt.imshow(u, cmap='hot', interpolation='nearest')
plt.colorbar()
plt.title(title)
plt.xlabel('x-axis')
plt.ylabel('y-axis')
plt.show()


# Parameters for the Poisson problem
grid_size = (50, 50)   # 50x50 grid points
boundary_conditions = {
    'top': 100,    # Top boundary condition (Dirichlet)
    'bottom': 0,   # Bottom boundary condition (Dirichlet)
    'left': 0,     # Left boundary condition (Dirichlet)
    'right': 0     # Right boundary condition (Dirichlet)
}
source_term = np.zeros((grid_size[0], grid_size[1]))   # No source
    term

# Solve the Poisson equation using finite difference
solution = solve_poisson_fd(grid_size, boundary_conditions,
    source_term)

# Plot the solution
plot_solution(solution, title='Solution to Poisson Equation using
    Finite Difference')
```

This code defines two main functions:

- `solve_poisson_fd` solves the Poisson equation using the finite difference method with specified boundary conditions and a source term.
- `plot_solution` visualizes the computed solution.

The main execution flow involves defining the grid size, boundary conditions, and source term, solving the Poisson equation using finite differences, and plotting the resulting temperature or potential distribution across the domain. The resulting plot gives a clear view of the solution spatially across the grid defined.

Multiple Choice Questions

1. What is a primary goal of the Dirichlet problem?

(a) To find solutions to PDEs with mixed boundary conditions
   (b) To find solutions to PDEs with prescribed values on a boundary
   (c) To approximate solutions using numerical methods
   (d) To establish existence and uniqueness theorems

2. Which of the following accurately describes the Dirichlet boundary condition?

   (a) The value of the solution is prescribed on the interior of the domain
   (b) The value of the derivative of the solution is prescribed on the boundary
   (c) The value of the solution is prescribed on the boundary of the domain
   (d) The integral of the solution is fixed over the boundary

3. In the context of weak solutions, which of the following is true about Sobolev spaces?

   (a) Functions in Sobolev spaces must be differentiable everywhere
   (b) Functions in Sobolev spaces may have jump discontinuities
   (c) Only continuous functions are included in Sobolev spaces
   (d) Sobolev spaces contain only polynomial functions

4. The variational formulation of the Dirichlet problem is equivalent to minimizing which functional?

   (a) The energy functional that represents total potential energy
   (b) The Dirichlet energy functional
   (c) The kinetic energy functional of the system
   (d) The work done by external forces

5. Finite element methods for solving the Dirichlet problem utilize which of the following characteristics?

   (a) They approximate solutions using global polynomials

(b) They discretize the domain into smaller, simpler pieces called elements

(c) They require analytic solutions to the PDE before application

(d) They convert the problem into a system of ordinary differential equations

6. Which of the following best describes a classical solution to the Dirichlet problem?

   (a) A solution that satisfies the PDE only in a weak sense

   (b) A solution that is piecewise continuous but not differentiable

   (c) A solution that is smooth and satisfies the PDE exactly

   (d) A solution that merely exists without giving specific values

7. In what type of engineering problems is the Dirichlet problem commonly applied?

   (a) Problems involving the dynamics of fluids and gases exclusively

   (b) Problems related to the static analysis of structures, heat distribution, and electromagnetic fields

   (c) Problems focusing solely on time-dependent phenomena

   (d) Problems where no boundary conditions are required

Answers:

1. **B: To find solutions to PDEs with prescribed values on a boundary** The primary goal of the Dirichlet problem is to find a solution to a partial differential equation (PDE) that satisfies specified values on the boundary of a given domain.

2. **C: The value of the solution is prescribed on the boundary of the domain** The Dirichlet boundary condition specifically involves prescribing the values of the solution function on the boundary, as opposed to interior conditions or derivatives.

3. **B: Functions in Sobolev spaces may have jump discontinuities** Sobolev spaces, by their nature, allow for functions that may not be smooth everywhere, including those with discontinuities, as long as their weak derivatives are square integrable.

4. **B: The Dirichlet energy functional** The variational formulation of the Dirichlet problem revolves around finding a function that minimizes the Dirichlet energy, thus establishing a link between potential energy minimization and the solutions to the PDE.

5. **B: They discretize the domain into smaller, simpler pieces called elements** Finite element methods utilize a mesh or partition into smaller elements, allowing for approximation of the solution within each element using polynomial functions, which facilitates the numerical solution of PDEs.

6. **C: A solution that is smooth and satisfies the PDE exactly** A classical solution to the Dirichlet problem refers to a solution that is sufficiently smooth and satisfies both the PDE and boundary conditions pointwise, distinguishing it from weak solutions.

7. **B: Problems related to the static analysis of structures, heat distribution, and electromagnetic fields** The Dirichlet problem is widely applicable in various engineering disciplines, especially in scenarios involving boundary value problems such as structural analysis, heat transfer, and electromagnetic field analysis.

Chapter 29

The Von Kármán Equations

Plate Theory in Structural Mechanics

The Von Kármán equations are a fundamental set of equations in plate theory, a branch of structural mechanics that deals with the behavior of thin plates subjected to various loads and boundary conditions. In this chapter, we explore the mathematical formulation and significance of the Von Kármán equations in structural analysis.

1 Assumptions and Governing Equations

The Von Kármán equations are derived using certain assumptions and simplifications that are appropriate for thin plates. These assumptions include:

- The plate is considered infinitesimally thin, with one dimension much smaller than the other two.

- The plate is relatively flat, with small deflections and rotations.

- The material of the plate is homogeneous and isotropic.

- Deformations of the plate are primarily due to bending and stretching, while shear deformations are neglected.

- The plate is subjected to external loads and boundary conditions, such as point forces, distributed loads, or prescribed displacements.

Under these assumptions, the governing equations for the Von Kármán equations can be derived from the principles of equilibrium and compatibility, resulting in a system of partial differential equations.

The primary equations in the Von Kármán equations are:

$$D_1\left(\frac{\partial^4 w}{\partial x^4} + 2\frac{\partial^4 w}{\partial x^2 \partial y^2} + \frac{\partial^4 w}{\partial y^4}\right) + D_2\nabla^2(\nabla^2 w) + D_3(\nabla^2 w)^2 = q(x,y) \tag{29.1}$$

where $w(x,y)$ represents the deflection of the thin plate in the vertical direction, $q(x,y)$ represents the applied load, and ∇^2 represents the Laplacian operator.

The coefficients D_1, D_2, and D_3 are related to the material properties of the plate and can be determined based on the specific problem at hand.

2 Boundary Conditions

The Von Kármán equations are typically accompanied by appropriate boundary conditions that specify the behavior of the thin plate at its edges. Common boundary conditions include:

- Simply Supported: The plate is supported along its edges and is free to rotate, but cannot transmit bending or shearing moments.

- Clamped: The plate is fully fixed along its edges, preventing any deflections or rotations.

- Free: The plate is completely free at its edges, allowing for both deflections and rotations.

The specific choice of boundary conditions depends on the physical system being analyzed and the desired behavior of the thin plate.

3 Significance in Structural Analysis

The Von Kármán equations play a crucial role in structural analysis, particularly for analyzing the behavior of thin plates under various loading conditions. These equations provide a mathematical framework for predicting the deflections and stresses in thin plates, allowing engineers to design structures that can withstand the applied loads while maintaining their integrity.

By solving the Von Kármán equations, engineers can obtain valuable insights into the bending and buckling behavior of thin plates, which is important in a wide range of applications, including aerospace engineering, civil engineering, and mechanical engineering.

The Von Kármán equations also serve as a foundation for more complex plate theories, such as the Mindlin-Reissner plate theory and the Kirchhoff-Love plate theory, which incorporate additional effects, such as transverse shear deformations and large rotations, respectively.

In summary, the Von Kármán equations provide a mathematical framework for analyzing the behavior of thin plates and are widely used in structural analysis and design. Their solution allows engineers to predict deflections, stresses, and buckling modes of thin plates subjected to various loading and boundary conditions.

Python Code Snippet

Below is a Python code snippet that implements the Von Kármán equations for analyzing the deflection of a thin plate subjected to uniform loading. The code uses finite difference methods to numerically solve the governing equations.

```python
import numpy as np
import matplotlib.pyplot as plt

def von_karman_plates(Lx, Ly, nx, ny, q, D, BC):
    '''
    Solve the Von Kármán equations for a thin plate.
    :param Lx: Length of the plate in x direction (meters).
    :param Ly: Length of the plate in y direction (meters).
    :param nx: Number of divisions in the x direction.
    :param ny: Number of divisions in the y direction.
    :param q: Uniform load applied (N/m^2).
    :param D: Plate rigidity parameter (Nm).
```

```python
    :param BC: Boundary conditions ('simply_supported', 'clamped',
    ↪    'free').
    :return: Deflection of the plate.
    '''
    # Initialize grid and parameters
    dx = Lx / (nx - 1)
    dy = Ly / (ny - 1)
    w = np.zeros((ny, nx))
    w_new = np.zeros_like(w)

    # Set boundary conditions
    apply_boundary_conditions(w, BC)

    # Finite difference iteration
    for iteration in range(10000):
        for i in range(1, ny-1):
            for j in range(1, nx-1):
                w_new[i, j] = (D * (
                    (w[i+1, j] - 2 * w[i, j] + w[i-1, j]) / dx**4 +
                    2 * (w[i+1, j+1] - 2 * w[i, j] + w[i-1, j-1]) /
                    ↪    (dx*dy)**2 +
                    (w[i, j+1] - 2 * w[i, j] + w[i, j-1]) / dy**4
                ) + q) / (D)

        # Update boundary conditions after each iteration
        apply_boundary_conditions(w_new, BC)

        # Check convergence
        if np.max(np.abs(w_new - w)) < 1e-6:
            break

        w = np.copy(w_new)

    return w

def apply_boundary_conditions(w, BC):
    '''
    Apply the specified boundary conditions to the plate deflection.
    :param w: Deflection array.
    :param BC: Boundary conditions.
    '''
    if BC == 'simply_supported':
        w[0, :] = 0   # Bottom edge
        w[-1, :] = 0  # Top edge
        w[:, 0] = 0   # Left edge
        w[:, -1] = 0  # Right edge
    elif BC == 'clamped':
        w[0, :] = 0   # Bottom edge
        w[-1, :] = 0  # Top edge
        w[:, 0] = 0   # Left edge
        w[:, -1] = 0  # Right edge
    elif BC == 'free':
        pass  # No boundary conditions applied
```

```
# Parameters for the analysis
Lx = 2.0     # Length of the plate in meters
Ly = 2.0     # Width of the plate in meters
nx = 21      # Number of divisions in x (plus one for boundary)
ny = 21      # Number of divisions in y (plus one for boundary)
q = 5000     # Uniform load (N/m^2)
D = 2.1e-5   # Rigidity of the plate (Nm)
BC = 'simply_supported'  # Boundary conditions type

# Calculate deflection
deflection = von_karman_plates(Lx, Ly, nx, ny, q, D, BC)

# Plotting the deflection
X, Y = np.meshgrid(np.linspace(0, Lx, nx), np.linspace(0, Ly, ny))
plt.figure(figsize=(10, 6))
plt.contourf(X, Y, deflection, levels=50, cmap='viridis')
plt.colorbar(label='Deflection (m)')
plt.title('Deflection of Thin Plate under Uniform Load')
plt.xlabel('Length (m)')
plt.ylabel('Width (m)')
plt.show()
```

This code defines several functions:

- `von_karman_plates` is responsible for setting up and solving the Von Kármán equations using finite difference methods, iterating until convergence is achieved.
- `apply_boundary_conditions` applies the specified boundary conditions (simply supported, clamped, or free) to the plate deflection.

In this implementation, the code simulates the deflection of a thin plate under uniform loading while allowing for visualization through contour plots.

Multiple Choice Questions

1. Which of the following assumptions is NOT made when deriving the Von Kármán equations?

    (a) The plate is infinitesimally thin.

    (b) Deformations due to shear are significant.

    (c) The plate is homogeneous and isotropic.

    (d) The plate is subjected to external loads.

2. Which physical variable represents the deflection of the plate in the Von Kármán equations?

(a) $q(x,y)$

(b) D_1

(c) $w(x,y)$

(d) ∇^2

3. What type of loading condition is addressed in the governing equations of the Von Kármán plates?

    (a) Distributed loads only

    (b) Concentrated point loads only

    (c) Both distributed loads and concentrated point loads

    (d) No external loads are considered

4. In the context of Von Kármán equations, what does the term D_1 generally represent?

    (a) The deflection of the plate

    (b) The material stiffness of the plate

    (c) The coefficient related to the Laplacian operator

    (d) The applied load on the plate

5. Which of the following boundary conditions allows for rotation but no translation at the edges of the plate?

    (a) Free boundary condition

    (b) Simply supported boundary condition

    (c) Clamped boundary condition

    (d) Fixed boundary condition

6. What physical theory do the Von Kármán equations primarily contribute to?

    (a) Fluid mechanics

    (b) Structural mechanics

    (c) Thermodynamics

    (d) Electromagnetism

7. When solving the Von Kármán equations, engineers seek to understand all of the following EXCEPT:

    (a) Stress distributions in the plate

(b) Predicted vibrations of the plate

(c) Plate deflections

(d) Buckling behavior under loads

Answers:
1. **B: Deformations due to shear are significant.** The Von Kármán equations assume that shear deformations are negligible, which is appropriate for thin plates.

2. **C:** $w(x,y)$ In the Von Kármán equations, $w(x,y)$ represents the vertical deflection of the plate, which is essential for analyzing plate behavior.

3. **C: Both distributed loads and concentrated point loads** The Von Kármán equations are capable of addressing a variety of loading conditions, including both distributed loads and concentrated point loads on the plate.

4. **B: The material stiffness of the plate** D_1 in the Von Kármán equations relates to the material properties and is a coefficient that expresses the stiffness of the plate in bending.

5. **B: Simply supported boundary condition** A simply supported boundary condition permits rotations without translation, allowing the edges to rotate freely while preventing vertical movements.

6. **B: Structural mechanics** The Von Kármán equations are a crucial part of structural mechanics, specifically in analyzing the behavior of plates under various loading conditions.

7. **B: Predicted vibrations of the plate** While stress distributions, deflections, and buckling behavior are all relevant outcomes studied in the context of the Von Kármán equations, predicted vibrations are not the primary focus of these equations, which mainly deal with static loads.

Chapter 30

Maximum Principles for Elliptic Equations

Statement of Maximum Principles

In this chapter, we explore the maximum principles for elliptic equations, which are powerful tools widely used in the analysis of partial differential equations. These principles provide crucial insights into the behavior and properties of solutions to elliptic equations, allowing for the derivation of important results such as uniqueness theorems, stability analysis, and the establishment of qualitative behavior.

1 The Strong Maximum Principle

The *strong maximum principle* is a fundamental result for elliptic equations, providing conditions under which a solution attains its maximum (or minimum) strictly *inside* the domain, rather than on the boundary. Let us consider the general second-order elliptic equation of the form:

$$Lu = -\sum_{i,j=1}^{n} \frac{\partial}{\partial x_i}\left(a_{ij}(x)\frac{\partial u}{\partial x_j}\right) + \sum_{i=1}^{n} b_i(x)\frac{\partial u}{\partial x_i} + c(x)u = 0, \quad (30.1)$$

where $x = (x_1, \ldots, x_n) \in \Omega \subset \mathbb{R}^n$ is the spatial variable, $u(x)$ is the unknown function to be solved, and L is a second-order linear elliptic operator. We state the strong maximum principle as follows:

Theorem (Strong Maximum Principle): Let u be a bounded solution of Equation (30.1) in a bounded domain Ω with smooth enough boundary $\partial \Omega$. If either of the following conditions holds:

1. $Lu \geq 0$ in Ω and u attains its maximum at an interior point of Ω, then u is a constant function.

2. $Lu \leq 0$ in Ω and u attains its minimum at an interior point of Ω, then u is a constant function.

then u is constant throughout Ω.

2 The Weak Maximum Principle

The *weak maximum principle* provides conditions under which the maximum (or minimum) of a solution to an elliptic equation is attained on the boundary of the domain. It is often more straightforward to apply than the strong maximum principle. We state the weak maximum principle as follows:

Theorem (Weak Maximum Principle): Let u be a bounded solution of Equation (30.1) in a bounded domain Ω with smooth enough boundary $\partial \Omega$. If either of the following conditions holds:

1. $Lu \geq 0$ in Ω and u satisfies $u \leq \max_{x \in \partial \Omega} u$,

2. $Lu \leq 0$ in Ω and u satisfies $u \geq \min_{x \in \partial \Omega} u$,

then u satisfies $u \leq \max_{x \in \partial \Omega} u$ or $u \geq \min_{x \in \partial \Omega} u$ respectively.

Proofs and Mathematical Insights

The proofs of the maximum principles rely on the properties of elliptic partial differential equations and the divergence theorem applied to suitable functions. They inherently exploit the non-negativity (or non-positivity) of the operator L, which is a fundamental characteristic of elliptic equations.

The strong maximum principle proof usually involves constructing a suitable function by exploiting either the non-negativity or non-positivity of the operator L. By using this function and applying the divergence theorem, it can be shown that the maximum (or minimum) of the solution u cannot be attained in the interior of the domain, leading to the desired conclusion of the constant function. The proof of the weak maximum principle follows a similar approach, but the choice of the test function may differ.

The maximum principles provide insights into the qualitative behavior of solutions. They establish that, under certain conditions, a solution to an elliptic equation cannot exhibit nontrivial local maxima or minima in the interior of the domain. This information is pivotal for understanding the uniqueness of solutions, stability properties, and the behavior of solutions near the boundary.

Additionally, the maximum principles have significant implications in mathematical modeling and scientific disciplines. They are extensively utilized in fields such as fluid dynamics, heat conduction, electrostatics, and diffusion processes, enabling researchers to analyze and predict the behavior of physical systems governed by elliptic equations.

Applications to Uniqueness Theorems

The maximum principles play a fundamental role in the establishment of uniqueness theorems for solutions of elliptic equations. By combining the maximum principles with suitable comparison arguments, mathematicians have derived powerful results that guarantee the uniqueness of solutions to various classes of elliptic equations.

For instance, the uniqueness of solutions to the Laplace's equation $\Delta u = 0$ subject to appropriate boundary conditions can be proven using the weak maximum principle. By showing that two solutions that satisfy the equation and boundary conditions must have the same maximum and minimum values, it follows that the solutions are identical.

Similarly, the maximum principles have been used to establish uniqueness results for more general elliptic equations, including nonlinear equations and systems of equations. By invoking the maximum principles and constructing appropriate comparison functions, mathematicians have shown that solutions to these equations are unique within certain classes of functions.

The uniqueness theorems derived from the maximum principles provide crucial mathematical foundations for constructing solutions and studying the behavior of elliptic equations. They ensure that, given suitable conditions, there is only one solution to a given problem, enhancing the reliability and predictive power of mathematical models.

Discrete Maximum Principles

In addition to the continuous maximum principles, there are discrete versions that apply to numerical approximations of elliptic equations. These discrete maximum principles are employed in the analysis of finite difference, finite element, and other numerical methods, providing important insights into the behavior and convergence of numerical solutions.

The discrete maximum principles are derived by introducing appropriate discrete counterparts of the continuous operators and establishing analogous inequalities. These principles enable researchers to verify the accuracy and stability of numerical schemes and assess their ability to preserve qualitative properties of the continuous solutions.

The discrete maximum principles have become indispensable tools in numerical analysis, guiding the design and implementation of numerical methods for elliptic equations. They help ensure the reliability and efficiency of numerical simulations in various fields, including computational physics, engineering, and applied mathematics.

Practical Applications

The maximum principles for elliptic equations find numerous practical applications in various scientific and engineering fields. Some notable applications include:

- **Heat Conduction:** The maximum principles are utilized to analyze and predict the behavior of temperature distributions in materials subjected to heat conduction. They allow for the determination of maximum and minimum temperatures within the domain and aid in the design of efficient cooling systems.

- **Fluid Dynamics:** In the context of fluid flow, the maximum principles play a crucial role in the analysis of velocity fields, pressure distributions, and vorticity in steady-state and time-dependent flows. They provide insights into the occurrence of stagnation points, hyperbolic and elliptic regions, and the presence of boundary layers.

- **Electrostatics:** The maximum principles are essential for studying electrostatic potential distributions in conductive and insulating materials. They allow for determining the location of maximum and minimum potential values, aiding in the design of electric field configurations and optimizing the performance of electrical devices.

- **Material Science:** The maximum principles are employed in the analysis of stress and strain distributions in elastic materials, aiding in the design and evaluation of structural components. They enable determining critical points, regions prone to failure, and the optimization of material properties.

These applications highlight the versatility and significance of the maximum principles in understanding and predicting the behavior of physical systems governed by elliptic equations. They contribute to the advancement of scientific knowledge, the development of new technologies, and the improvement of engineering designs.

Conclusion

In this chapter, we have explored the maximum principles for elliptic equations, including the strong and weak maximum principles. These principles provide crucial insights into the behavior, properties, and uniqueness of solutions to elliptic equations. The maximum principles have wide-ranging applications in various scientific and engineering fields and are instrumental in the analysis and prediction of physical phenomena. Moreover, the discrete maximum principles play a vital role in the analysis of numerical methods for solving elliptic equations. These principles ensure the reliability and accuracy of numerical approximations, facilitating the numerical simulation of diverse systems. Overall, the maximum principles offer valuable mathematical tools that enhance our understanding of elliptic equations and their applications in practical settings.

Python Code Snippet

Below is a comprehensive Python code snippet that implements the Strong Maximum Principle and provides numerical solutions to a second-order elliptic partial differential equation, utilizing finite difference methods.

```python
import numpy as np
import matplotlib.pyplot as plt

def laplace_solver(domain_size, grid_points, boundary_func,
                   max_iter=10000, tol=1e-5):
    '''
    Solves the Laplace equation using the finite difference method.

    :param domain_size: Size of the domain (Lx, Ly).
    :param grid_points: Number of grid points (Nx, Ny).
    :param boundary_func: Function defining boundary conditions.
    :param max_iter: Maximum number of iterations.
    :param tol: Convergence tolerance.
    :return: 2D array of computed values.
    '''
    # Create grid
    Lx, Ly = domain_size
    Nx, Ny = grid_points
    x = np.linspace(0, Lx, Nx)
    y = np.linspace(0, Ly, Ny)
    u = np.zeros((Nx, Ny))  # Initialize potential array

    # Set boundary conditions
    u[:, 0] = boundary_func(x, 0)       # Bottom boundary
    u[:, -1] = boundary_func(x, Ly)     # Top boundary
    u[0, :] = boundary_func(0, y)       # Left boundary
    u[-1, :] = boundary_func(Lx, y)     # Right boundary

    # Iterative solver
    for iter_count in range(max_iter):
        u_old = u.copy()
        u[1:-1, 1:-1] = 0.25 * (u_old[1:-1, :-2] + u_old[1:-1, 2:] +
                                u_old[:-2, 1:-1] + u_old[2:, 1:-1])
        # Check for convergence
        if np.max(np.abs(u - u_old)) < tol:
            print(f"Converged in {iter_count} iterations.")
            break
    return u

def boundary_condition(x, y):
    '''
    Define boundary conditions for the Laplace equation.
    :param x: x-coordinates.
    :param y: y-coordinates.
```

```
    :return: Boundary values.
    '''
    # Example: u = sin(pi*x)*sin(pi*y) on boundary
    return np.sin(np.pi * x) * np.sin(np.pi * y)

# Parameters
domain_size = (1.0, 1.0)        # Domain dimensions (Lx, Ly)
grid_points = (50, 50)          # Number of grid points (Nx, Ny)

# Solve the Laplace equation
solution = laplace_solver(domain_size, grid_points,
↪   boundary_condition)

# Plotting the results
X = np.linspace(0, domain_size[0], grid_points[0])
Y = np.linspace(0, domain_size[1], grid_points[1])
X, Y = np.meshgrid(X, Y)

plt.figure(figsize=(8, 6))
plt.contourf(X, Y, solution.T, levels=50, cmap='viridis')
plt.colorbar(label='Potential u(x,y)')
plt.title('Solution of Laplace Equation')
plt.xlabel('x')
plt.ylabel('y')
plt.grid()
plt.show()
```

This code defines a function called `laplace_solver` that implements a finite difference method to solve the Laplace equation in a two-dimensional domain given specific boundary conditions, defined by the function `boundary_condition`.

1. **laplace_solver**: This function initializes a grid, applies boundary conditions, and iteratively updates the potential values until they converge based on a specified tolerance.

2. **boundary_condition**: This function defines the boundary conditions used in the simulation.

At the end of the execution, the computed potential distribution in the domain is visualized using the Matplotlib library, displaying a contour plot of the potential values.

The provided example solves the Laplace equation for a square domain and should work effectively for visualization of potential distributions under the specified boundary conditions.

Multiple Choice Questions

1. What does the strong maximum principle state about the behavior of a bounded solution u of a second-order elliptic equation in a bounded domain?

    (a) u can have its maximum in the interior of the domain without being constant.

    (b) If $Lu \geq 0$ and u has a maximum in Ω, then u must be constant.

    (c) u must attain its maximum on the boundary of the domain.

    (d) There are no restrictions on where u can attain its maximum.

2. For which of the following conditions does the weak maximum principle apply?

    (a) $Lu \leq 0$ in Ω and u has a maximum at an interior point.

    (b) $Lu \geq 0$ in Ω and u attains its minimum at an interior point.

    (c) $Lu \geq 0$ in Ω and u is less than or equal to its maximum on the boundary.

    (d) $Lu = 0$ in Ω.

3. Which approach is typically used in the proof of the strong maximum principle?

    (a) Direct integration of the solution u.

    (b) Constructing a suitable test function and applying the divergence theorem.

    (c) Employing numerical approximations of the solution.

    (d) Analyzing the Fourier series expansion of u.

4. How do maximum principles contribute to the uniqueness of solutions for elliptic equations?

    (a) They guarantee that all solutions are unique.

    (b) They allow for constructing comparison functions that show two solutions cannot overlap.

    (c) They state that solutions must be continuous.

(d) They imply that solutions can only exist under very specific conditions.

5. In which of the following fields are maximum principles particularly applied?

    (a) Number theory.
    (b) Quantum computing.
    (c) Fluid dynamics.
    (d) Abstract algebra.

6. True or False: The discrete maximum principle guarantees that the numerical solutions of elliptic equations mirror the behavior of the continuous solutions.

    (a) True
    (b) False

7. The application of the maximum principles allows researchers to determine:

    (a) The total number of solutions to elliptic equations.
    (b) The locality of extrema of solutions.
    (c) The exact form of the boundary conditions.
    (d) The singularities in the solution space.

Answers:

1. **B: If $Lu \geq 0$ and u has a maximum in Ω, then u must be constant.** The strong maximum principle asserts that if a bounded solution to an elliptic equation attains its maximum inside the domain, then it is constant throughout the domain.

2. **C: $Lu \geq 0$ in Ω and u is less than or equal to its maximum on the boundary.** The weak maximum principle applies when the solution does not exceed the maximum value on the boundary and provides greater flexibility than the strong maximum principle regarding where extrema can occur.

3. **B: Constructing a suitable test function and applying the divergence theorem.** The proofs of maximum principles rely on constructing specific test functions that leverage the properties of the elliptic operator and the divergence theorem to show extremal behavior.

4. **B: They allow for constructing comparison functions that show two solutions cannot overlap.** Maximum principles

are instrumental in uniqueness theorems because if two solutions were to share a common maximum or minimum, the principles can demonstrate that they must be the same under suitable conditions.

5. **C: Fluid dynamics.** Maximum principles have significant applications in fluid dynamics, where they help analyze velocity fields and pressure distributions.

6. **A: True** The discrete maximum principle is designed to ensure that discrete numerical solutions also exhibit maximum and minimum behaviors analogous to their continuous counterparts.

7. **B: The locality of extrema of solutions.** The maximum principles help determine where local maxima and minima occur in the domain of the solution, providing crucial insights into the solution's behavior.

Chapter 31

Nonlinear Elliptic Equations

Introduction and Examples

Nonlinear elliptic equations are a fascinating area of study in the field of partial differential equations (PDEs). Unlike linear elliptic equations, where the highest-order derivatives of the unknown function appear linearly, nonlinear elliptic equations involve nonlinear terms. This nonlinearity introduces complexities and challenges in both theoretical analysis and numerical approximation.

Nonlinear elliptic equations appear in various branches of mathematics, physics, and engineering. They find applications in areas such as fluid mechanics, mathematical biology, materials science, and geometric analysis, to name a few. The equations encompass a diverse range of mathematical models, including reaction-diffusion systems, minimal surface equations, and equations with critical or supercritical growth. Studying and understanding these equations is crucial for gaining insights into the behavior of physical and mathematical phenomena.

In this chapter, we explore the theory and properties of nonlinear elliptic equations. We focus on second-order quasilinear equations of the form:

$$Lu = \text{div}(A(x, u, \nabla u)) + G(x, u, \nabla u) = 0, \qquad (31.1)$$

where $x \in \Omega \subset \mathbb{R}^n$ is the spatial variable, u is the unknown function, A is a nonlinear function representing the coefficient matrix, and G encompasses the nonlinear terms. We delve into the existence and regularity of solutions, uniqueness theorems, and qualitative properties of these equations.

1 Examples of Nonlinear Elliptic Equations

To facilitate understanding, we present examples of nonlinear elliptic equations that arise in various applications.

Example 1: The Allen-Cahn Equation

The Allen-Cahn equation is a well-known nonlinear elliptic equation that arises in materials science, phase transitions, and pattern formation. It can be written as:

$$\Delta u - u + u^3 = 0. \tag{31.2}$$

This equation models the phase separation process in materials undergoing a phase transition.

Example 2: The Lane-Emden Equation

The Lane-Emden equation is a nonlinear elliptic equation that arises in astrophysics and describes the density distribution of a polytropic gas sphere undergoing gravitational collapse. It can be written as:

$$\Delta u + u^n = 0, \tag{31.3}$$

where n is a positive constant.

Example 3: Reaction-Diffusion Equations

Reaction-diffusion equations form another important class of nonlinear elliptic equations. They model the diffusion of chemicals and the interaction between species in biological systems. An example is the Fisher-KPP equation:

$$\Delta u - \alpha u + \beta u^2 = 0, \tag{31.4}$$

where α and β are positive constants.

These examples highlight the diversity and significance of nonlinear elliptic equations in modeling a wide range of phenomena. They provide a starting point for exploring the properties and behavior of solutions to these equations.

Sobolev Spaces for Nonlinear Problems

Sobolev spaces provide a crucial framework for studying nonlinear elliptic equations. These function spaces characterize the differentiability and integrability properties of functions. In the context of nonlinear equations, Sobolev spaces allow us to define suitable function spaces on which to seek solutions.

1 Definition of Sobolev Spaces

We begin by defining the Sobolev space $W^{1,p}(\Omega)$, which consists of functions u whose distributional derivatives $\partial u/\partial x_i$ are square integrable on Ω, together with u itself being square integrable. This is given by:

$$W^{1,p}(\Omega) = \left\{ u \in L^p(\Omega) \,\middle|\, \frac{\partial u}{\partial x_i} \in L^p(\Omega) \text{ for } i = 1, \ldots, n \right\}, \quad (31.5)$$

where p is a real number such that $1 \leq p < \infty$, $L^p(\Omega)$ denotes the Lebesgue space of functions whose p-th power is integrable over Ω, and n is the spatial dimension.

For nonlinear elliptic equations, we often require higher regularity, and thus, we introduce the Sobolev space $W_0^{1,p}(\Omega)$ as the closure of smooth functions with compact support in $W^{1,p}(\Omega)$ with respect to the norm induced by $W^{1,p}(\Omega)$. In other words:

$$W_0^{1,p}(\Omega) = \overline{C_0^\infty(\Omega)}^{W^{1,p}(\Omega)}, \quad (31.6)$$

where $C_0^\infty(\Omega)$ denotes the space of smooth functions with compact support in Ω.

These Sobolev spaces provide the appropriate function spaces in which to pose and seek solutions to nonlinear elliptic equations. They ensure the required regularity for the derivatives involved in the equation and facilitate the formulation of suitable variational formulations.

2 Variational Formulations

Variational formulations play a crucial role in the study of nonlinear elliptic equations, as they allow us to characterize solutions through the minimization of appropriate energy functionals. For example, the weak form of a general nonlinear elliptic equation can be written as finding $u \in W_0^{1,p}(\Omega)$ such that:

$$\int_\Omega A(x, u, \nabla u) \cdot \nabla \varphi + G(x, u, \nabla u)\varphi \, dx = 0, \qquad (31.7)$$

for all $\varphi \in W_0^{1,p}(\Omega)$.

By formulating the equation in this variational manner, we can exploit the properties of Sobolev spaces and derive existence and uniqueness results for solutions. These variational techniques also enable us to extend the analysis to more general classes of nonlinear elliptic equations beyond quasilinear equations.

Existence Theorems

Establishing the existence of solutions to nonlinear elliptic equations is a fundamental result in the theory of these equations. In many cases, proving the existence of a solution requires utilizing both the variational techniques and theorems from nonlinear analysis.

For quasilinear equations, under suitable conditions on the coefficient matrix A and the nonlinear terms G, we can use the celebrated Leray-Schauder degree theory to establish the existence of solutions. The Leray-Schauder degree is a topological concept that encodes the mapping properties of the nonlinear operator involved in the equation.

When dealing with fully nonlinear equations, such as equations with critical or supercritical growth, existence results often rely on methods from critical point theory, variational methods, and the theory of convex sets.

The existence theorems for nonlinear elliptic equations contribute to our understanding of the behavior and properties of solutions. They provide a solid foundation for the further analysis and study of these equations.

Regularity and Stability

The regularity of solutions to nonlinear elliptic equations is another important aspect to investigate. It concerns the smoothness properties of solutions and their behavior near singular points or on the boundary of the domain.

For quasilinear equations, under suitable assumptions on the coefficient matrix A and the nonlinear terms G, regularity results can be established using techniques such as Schauder estimates and bootstrapping arguments. These results provide information about the smoothness of solutions, their higher derivatives, and the validity of classical solutions.

In the case of fully nonlinear equations, regularity results are more challenging to establish. They often require advanced techniques from partial differential equations, harmonic analysis, and geometric measure theory. The regularity theory provides insights into the behavior of solutions and enables the study of stability, convergence of numerical methods, and the interplay between regularity and the geometry of the domain.

Stability analysis is crucial in understanding the robustness and sensitivity of solutions to perturbations in the equation or the domain. Stability results for nonlinear elliptic equations provide insights into the continuous dependence of solutions on the data and the qualitative properties of solutions under small changes.

1 Regularity and Stability of Solutions

The regularity and stability of solutions to nonlinear elliptic equations depend on the smoothness of the coefficient matrix A, the growth conditions on the nonlinear terms G, and the geometric properties of the domain Ω. The regularity theory enables us to establish the Hölder continuity and differentiability of solutions, providing a deeper understanding of their behavior.

Stability results reveal the robustness of solutions under perturbations and small variations. By quantifying the sensitivity of solutions to changes in the equation or the domain, stability analysis aids in the verification of numerical methods, the design of experiments, and the understanding of the long-term behavior of solutions.

Applications in Nonlinear Analysis

Nonlinear elliptic equations have applications in various fields of nonlinear analysis. They serve as prototypes for studying the behavior of solutions to more general classes of nonlinear partial differential equations and systems.

Nonlinear analysis utilizes methods and techniques from functional analysis, convex analysis, fixed point theory, and variational methods to investigate the properties and structure of solutions to nonlinear equations. Understanding the dynamics of solutions, bifurcation phenomena, and the geometry of solution sets are essential topics in nonlinear analysis, of which the theory of nonlinear elliptic equations is a cornerstone.

Moreover, nonlinear elliptic equations play a central role in the study of problems in calculus of variations, shape optimization, and free boundary problems. They find applications in diverse fields such as mathematical physics, mathematical biology, geometric analysis, and materials science. The study of nonlinear elliptic equations has yielded valuable insights into the behavior of physical and mathematical phenomena.

Conclusion

In this chapter, we have explored the theory and properties of nonlinear elliptic equations. These equations constitute a rich and diverse class of partial differential equations with applications in various scientific and engineering fields. By focusing on second-order quasilinear equations, we have discussed the importance of Sobolev spaces in formulating appropriate variational formulations and establishing existence and regularity results. Additionally, we have highlighted the significance of stability analysis and its implications in understanding the behavior of solutions under perturbations. The study of nonlinear elliptic equations contributes not only to the understanding of specific models but also to the broader field of nonlinear analysis, providing insights into the properties of general nonlinear equations and systems.

Python Code Snippet

Below is a Python code snippet that implements important algorithms and calculations related to nonlinear elliptic equations discussed in this chapter, including the Allen-Cahn equation, the Lane-Emden equation, and methods for solving nonlinear PDEs using finite difference and variational formulations.

```
import numpy as np
import matplotlib.pyplot as plt
```

```python
def allen_cahn_equation(u, num_iterations, dt, epsilon):
    '''
    Solve the Allen-Cahn equation using an explicit finite
    ↪ difference scheme.
    :param u: Initial condition as a 2D numpy array.
    :param num_iterations: Number of time iterations.
    :param dt: Time step size.
    :param epsilon: Parameter controlling the interface width.
    :return: Updated u after num_iterations.
    '''
    for _ in range(num_iterations):
        laplacian = (np.roll(u, 1, axis=0) + np.roll(u, -1, axis=0)
    ↪           +
                     np.roll(u, 1, axis=1) + np.roll(u, -1, axis=1)
    ↪           - 4*u)
        u += dt * (epsilon**2 * laplacian - u + u**3)
    return u

def lane_emden_equation(n, num_points, x_max):
    '''
    Solve the Lane-Emden equation using the finite difference
    ↪ method.
    :param n: Polytropic index.
    :param num_points: Number of discretization points.
    :param x_max: Maximum x value for the domain.
    :return: x values and corresponding u values as lists.
    '''
    x = np.linspace(0, x_max, num_points)
    u = np.zeros(num_points)
    u[0] = 1  # Initial condition
    dx = x[1] - x[0]

    for i in range(1, num_points-1):
        u[i+1] = (2 * u[i] - u[i-1] + dx**2 * (-u[i]**n)) / (1 +
    ↪       dx**2)

    return x, u

def variational_solver(A, b):
    '''
    Solve a linear system Ax = b using numpy's linear algebra
    ↪ solver.
    :param A: Coefficient matrix.
    :param b: Right-hand side vector.
    :return: Solution x.
    '''
    return np.linalg.solve(A, b)

# Parameters for Allen-Cahn
grid_size = 100
u_initial = np.random.rand(grid_size, grid_size) * 2 - 1  # Random
↪   initial condition
```

```
num_iterations = 1000
dt = 0.01
epsilon = 0.01

# Solve Allen-Cahn equation
u_result = allen_cahn_equation(u_initial, num_iterations, dt,
↪ epsilon)

# Parameters for Lane-Emden
n = 3    # Polytropic index
num_points = 100
x_max = 10

# Solve Lane-Emden equation
x_lan, u_lan = lane_emden_equation(n, num_points, x_max)

# Visualize results for Allen-Cahn
plt.figure(figsize=(12, 6))
plt.subplot(1, 2, 1)
plt.imshow(u_result, cmap='gray')
plt.title('Allen-Cahn Equation Solution')
plt.colorbar()

# Visualize results for Lane-Emden
plt.subplot(1, 2, 2)
plt.plot(x_lan, u_lan)
plt.title('Lane-Emden Equation Solution')
plt.xlabel('x')
plt.ylabel('u(x)')
plt.grid()

plt.tight_layout()
plt.show()
```

This code defines three functions:

- `allen_cahn_equation` solves the Allen-Cahn equation using an explicit finite difference scheme.
- `lane_emden_equation` provides a numerical solution for the Lane-Emden equation using the finite difference method.
- `variational_solver` solves a linear system $Ax = b$ using NumPy's linear algebra capabilities.

Additionally, it visualizes the results of both equations, showcasing the solution space for the Allen-Cahn equation and the behavior of the Lane-Emden equation across a specified domain.

Multiple Choice Questions

1. Which of the following types of equations does the Allen-Cahn equation represent?

    (a) Linear elliptic equation

    (b) Nonlinear elliptic equation

    (c) Hyperbolic equation

    (d) Parabolic equation

2. What is the primary use of Sobolev spaces in the context of nonlinear elliptic equations?

    (a) To define functions that have an infinite number of derivatives

    (b) To provide a framework for variational formulations and regularity results

    (c) To solve ordinary differential equations

    (d) To approximate solutions of linear equations

3. Which theorem is commonly used to establish the existence of solutions for nonlinear elliptic equations?

    (a) Banach Fixed-Point Theorem

    (b) Leray-Schauder Degree Theory

    (c) Cauchy-Schwarz Inequality

    (d) Sturm-Liouville Theory

4. What is the main goal of stability analysis in the study of nonlinear elliptic equations?

    (a) To prove the uniqueness of solutions

    (b) To analyze the robustness of solutions against perturbations

    (c) To derive numerical approximations

    (d) To find the minimum value of the solutions

5. In which application area do nonlinear elliptic equations NOT typically play a role?

    (a) Fluid mechanics

(b) Astrophysics

(c) Quantum mechanics

(d) Medical imaging

6. The Lane-Emden equation includes a term that accounts for:

    (a) Linear growth

    (b) Nonlinear growth

    (c) Constant growth

    (d) Exponential decay

7. Which of the following statements about variational formulations of nonlinear elliptic equations is TRUE?

    (a) They do not require boundary conditions.

    (b) They involve minimizing energy functionals associated with the equations.

    (c) They can only be applied to linear equations.

    (d) Solutions are defined pointwise at every point in the domain.

Answers:

1. **B: Nonlinear elliptic equation** The Allen-Cahn equation is a classic example of a nonlinear elliptic equation, specifically used in the study of phase transitions and pattern formation in materials.

2. **B: To provide a framework for variational formulations and regularity results** Sobolev spaces are essential for establishing the existence, uniqueness, and regularity of solutions to nonlinear elliptic equations due to their integrability and differentiability properties.

3. **B: Leray-Schauder Degree Theory** The Leray-Schauder degree theory is a topological tool used to prove the existence of solutions for a range of nonlinear elliptic equations, particularly in the setting of quasilinear equations.

4. **B: To analyze the robustness of solutions against perturbations** Stability analysis helps determine how sensitive solutions are to changes in parameters or boundary conditions, providing insights into the practical implications of the solutions.

5. **D: Medical imaging** Nonlinear elliptic equations are primarily used in fields like fluid mechanics, astrophysics, and quantum mechanics, but are not typically associated with medical imaging.

6. **B: Nonlinear growth** The Lane-Emden equation features a term that introduces nonlinear growth, which is essential for modeling the self-gravitation of polytropic gas spheres.

7. **B: They involve minimizing energy functionals associated with the equations** Variational formulations of nonlinear elliptic equations are based on minimization principles, allowing for the characterization of solutions in terms of energy functionals. This approach is fundamental for deriving existence and regularity results.

Chapter 32

The De Giorgi-Nash-Moser Theorem

The De Giorgi-Nash-Moser theorem is a fundamental result in the theory of nonlinear elliptic equations. It provides a powerful tool for establishing regularity and other qualitative properties of solutions. In this chapter, we explore the theorem's significance, statement, and implications in the context of partial differential equations. We discuss its applications in various areas of mathematics and the insights it offers into the behavior of solutions to nonlinear elliptic equations.

Background and Importance

The De Giorgi-Nash-Moser theorem stems from the classic problem of understanding the regularity of solutions to elliptic equations. For linear elliptic equations, the regularity theory is well-established. However, extending this theory to nonlinear elliptic equations presents significant challenges due to the presence of nonlinear terms and the potential lack of maximum principles.

The theorem addresses these challenges by providing conditions under which solutions to a wide class of nonlinear elliptic equations exhibit improved regularity. It demonstrates that solutions to certain classes of equations are not only continuous but also possess higher degrees of smoothness.

The importance of the De Giorgi-Nash-Moser theorem lies in its far-reaching implications. It has been a key component in the development of the regularity theory for nonlinear partial differential equations. The theorem's techniques and ideas have influenced a wide range of research in the field and have applications in areas such as geometric analysis, mathematical physics, and harmonic analysis.

Statement of the Theorem

The De Giorgi-Nash-Moser theorem provides conditions under which solutions to nonlinear elliptic equations attain higher regularity. While the precise statement varies depending on the specific class of equations, the general principle remains consistent.

1 Monotonicity and Localization Properties

The theorem establishes monotonicity and localization properties of solutions. It states that if a suitable quantity associated with the solution is monotonic, then solutions are locally bounded and smoothly concentrated in compact sets.

For example, consider the following general form of a quasilinear elliptic equation:

$$Lu = \sum_{i,j=1}^{n} a_{ij}(x, u, \nabla u) u_{x_i x_j} + G(x, u, \nabla u) = 0,$$

where L is a quasilinear differential operator, a_{ij} and G are suitably smooth functions representing the coefficient matrix and the nonlinear terms, respectively.

The De Giorgi-Nash-Moser theorem establishes that under appropriate conditions, if u_1 and u_2 are two solutions to the above equation that satisfy $u_1 \leq u_2$, then there exists a compact set $K \subset \Omega$ such that u_1 and u_2 are locally bounded and concentrated on K.

2 Existence of Suitable Integrals

To establish the desired regularity, the theorem requires the existence of certain integral functionals associated with the solution.

These functionals, known as Moser's type integrals, play a crucial role in the proof of the theorem.

The Moser's type integrals capture the oscillatory behavior of the solution and allow for the control of its growth. By constructing suitable integral functionals and demonstrating their favorable properties, the theorem ensures that the solutions exhibit higher degrees of regularity.

The precise conditions and requirements for the existence and properties of these integrals vary depending on the specific class of equations and the associated structure of the coefficient matrix and the nonlinear terms.

3 Generalization to Nonlinear Systems and Higher Dimensions

The De Giorgi-Nash-Moser theorem can also be extended to nonlinear systems of elliptic equations and higher dimensions. The statement and techniques adapt accordingly to accommodate the additional complexity introduced by these generalizations.

In the case of systems, the monotonicity assumption is extended component-wise, and the localization properties hold for each component independently. The existence of suitable integral functionals is established for each component of the system, ensuring the desired regularity results for the entire system.

In higher dimensions, increased technical difficulties arise due to the additional complexity of the geometry. Despite these challenges, the theorem's fundamental principles remain applicable, leading to insights into the regularity of solutions in higher-dimensional settings.

Regularity Theory and Applications

The De Giorgi-Nash-Moser theorem forms a cornerstone of the regularity theory for nonlinear elliptic equations. It provides crucial tools and techniques for analyzing and establishing higher regularity properties of solutions.

Regularity theory enables the understanding of various aspects of the solutions, including their smoothness, differentiability, and the relationship between regularity and the equation's structure. It plays a vital role in studying qualitative properties of solutions, such as stability, uniqueness, and the interplay between regularity and boundary behavior.

The applications of the De Giorgi-Nash-Moser theorem extend beyond the realm of partial differential equations. Its techniques have influenced diverse areas of mathematics, such as geometric analysis, mathematical physics, harmonic analysis, and calculus of variations.

The theorem's implications can be seen in the study of geometric problems, where it bridges the gap between geometric features and the regularity of associated equations. Additionally, its ideas have led to advancements in the analysis of free boundaries, obstacle problems, and phase transitions.

Furthermore, the theorem has inspired the development of various refined tools, such as the Harnack inequality and the Krylov-Safonov theorem, which further contribute to the study of nonlinear elliptic equations and related areas.

In summary, the De Giorgi-Nash-Moser theorem is a fundamental result in the theory of nonlinear elliptic equations. It establishes conditions under which solutions exhibit higher regularity and provides powerful techniques for analyzing and understanding solutions in nonlinear settings. The theorem's applications extend beyond the realm of partial differential equations, influencing research in various areas of mathematics and offering insights into the behavior of solutions in complex systems.

Python Code Snippet

Below is a Python code snippet that implements numerical techniques for solving elliptic partial differential equations, specifically focusing on the nonlinear elliptic equation discussed in this chapter and applying the principles of the De Giorgi-Nash-Moser theorem.

```python
import numpy as np
import matplotlib.pyplot as plt

def solve_quasilinear_elliptic(L, u_initial, max_iter=1000,
    tol=1e-5):
    '''
    Solve a quasilinear elliptic equation using the fixed-point
        iteration method.
    :param L: A function representing the quasilinear operator.
    :param u_initial: Initial guess for the solution.
    :param max_iter: Maximum number of iterations.
```

```python
    :param tol: Tolerance for convergence.
    :return: Approximated solution.
    '''
    u = u_initial.copy()
    for i in range(max_iter):
        u_new = L(u)  # Apply the operator
        if np.max(np.abs(u_new - u)) < tol:
            print(f'Converged in {i} iterations.')
            return u_new
        u = u_new
    print('Max iterations reached. Convergence not achieved.')
    return u

def quasilinear_operator(u):
    '''
    Define the quasilinear operator for the elliptic equation.
    :param u: Current solution guess.
    :return: Updated solution based on the operator.
    '''
    n = u.shape[0]
    u_new = np.zeros_like(u)
    for i in range(1, n-1):
        for j in range(1, n-1):
            # Example update rule; here we are using a simple finite
            ↪   difference scheme
            u_new[i, j] = 0.25 * (u[i+1, j] + u[i-1, j] + u[i, j+1]
            ↪   + u[i, j-1]) - 0.1 * u[i, j]**2
    return u_new

# Set up the numerical grid
grid_size = 50
u_initial = np.random.rand(grid_size, grid_size)

# Solve the equation
solution = solve_quasilinear_elliptic(quasilinear_operator,
↪   u_initial)

# Plotting the solution
plt.imshow(solution, cmap='hot', interpolation='nearest')
plt.colorbar()
plt.title('Solution of the Nonlinear Elliptic Equation')
plt.show()
```

This code defines two main functions:

- `solve_quasilinear_elliptic` implements a simple fixed-point iteration method to solve a quasilinear elliptic equation iteratively.
- `quasilinear_operator` defines the specific form of the nonlinear operator applied in this example, representing a simple form of the finite difference scheme used in elliptic equations.

The provided example initializes a random grid, then solves the nonlinear elliptic equation using the defined methods and finally visualizes the solution with a heatmap. This approach exemplifies the techniques relevant to the regularity theory and application areas discussed in the chapter.

Multiple Choice Questions

1. What does the De Giorgi-Nash-Moser theorem primarily address?

    (a) The uniqueness of solutions to elliptic equations

    (b) The regularity of solutions to nonlinear elliptic equations

    (c) The existence of solutions to linear elliptic equations

    (d) The stability of solutions to parabolic equations

2. In the context of the De Giorgi-Nash-Moser theorem, what plays a crucial role in establishing higher regularity?

    (a) Integral functionals of a certain type

    (b) Maximum principles

    (c) Local existence results

    (d) Boundary conditions

3. Which of the following is NOT a requirement for applying the De Giorgi-Nash-Moser theorem?

    (a) Monotonicity of the solution

    (b) Local boundedness of the solution

    (c) Existence of suitable integral functionals

    (d) Smoothness of the coefficient matrix

4. The Moser's type integrals relate to what aspect of solutions in the De Giorgi-Nash-Moser theorem?

    (a) Nonlinear terms

    (b) Oscillatory behavior

    (c) Boundary conditions

    (d) Uniqueness

5. The theorem allows for its application to which kind of systems?

    (a) Linear systems only
    (b) Nonlinear systems in one dimension
    (c) Nonlinear systems and higher-dimensional cases
    (d) Systems with constant coefficients only

6. What kind of equations does the De Giorgi-Nash-Moser theorem specifically address?

    (a) Systems of parabolic equations
    (b) Linear elliptic equations
    (c) Nonlinear elliptic equations
    (d) Ordinary differential equations

7. True or False: The De Giorgi-Nash-Moser theorem can be used to establish regularity for solutions to any type of partial differential equation.

    (a) True
    (b) False

Answers:
1. **B: The regularity of solutions to nonlinear elliptic equations** The De Giorgi-Nash-Moser theorem specifically focuses on establishing conditions under which solutions to nonlinear elliptic equations exhibit higher regularity properties, thus enhancing our understanding of their behavior.

2. **A: Integral functionals of a certain type** The existence of suitable integral functionals, known as Moser's type integrals, is crucial for demonstrating the improved regularity of solutions as they capture the oscillatory behavior and control the growth of solutions.

3. **D: Smoothness of the coefficient matrix** While monotonicity, local boundedness, and the existence of integral functionals are required, the smoothness of the coefficient matrix itself is not an explicit requirement of the theorem.

4. **B: Oscillatory behavior** Moser's type integrals are specifically designed to capture the oscillatory behavior of solutions, enabling the theorem to show that these solutions have improved regularity properties.

5. **C: Nonlinear systems and higher-dimensional cases**
The theorem can be generalized to nonlinear systems of equations and is applicable in higher dimensions, indicating that its principles extend beyond single equations to more complex scenarios.

6. **C: Nonlinear elliptic equations** The De Giorgi-Nash-Moser theorem pertains to nonlinear elliptic equations, exploring the conditions under which solutions attain higher regularity.

7. **B: False** The De Giorgi-Nash-Moser theorem is not universally applicable to all types of partial differential equations; it is specifically tailored for certain classes of nonlinear elliptic equations, thereby having limitations on its applicability.

www.ingramcontent.com/pod-product-compliance
Lightning Source LLC
Chambersburg PA
CBHW071913210526
45479CB00002B/396